THE BACK H

SERIES TITLES

This Season, The Next
Casey Knott

Lessons in Geography
Phillip Sterling

Points of Tangency: Essays
Scott Russell Morris

Wildlifer
Neil F. Payne

We Come from Good Stock
Kay Oakes Oring

Squatter
Yolanda DeLoach

The Arc of the Escarpment
Robert Root

Soul of the Outdoors
Dave Greschner

From the Heart: The Story of Matrix
John Harmon

The Long Fields
Anne-Marie Oomen

Kick Out the Bottom
Erik Mortenson & Christopher Kramer

Wrong Tree: Adventures in Wildlife Biology
Jeff Wilson

At the Lake
Jim Landwehr

Body Talk
Takwa Gordon

The In-Between State
Martha Lundin

North Freedom
Carolyn Dallmann

Ohio Apertures
Robert Miltner

"Casey Knott is talking about a ton more than chickens when she writes, 'how does one explain to a bird that caging them is for their own good when really it's for my peace of mind?' In *This Season, The Next*, Knott makes a home, makes a family—the verb make here being the opposite of to ponder, the opposite of to wallow in, the making a re-making, a making out of loss, out of loneliness, a single mother post divorce doing things DIY because that's all she's got. With a newly blended family in a strangely rural acreage in the middle of urban Des Moines—a place that needs TLC but has the bones to become a paradise—Knott and her new husband, who met via a dating app, teach a next generation about what you can control and what you can't as they are learning it themselves. 'The things that mark us are small, yet enveloping,' writes Knott, and the best lessons here are found in these subtle metaphors that anchor the action like roots we didn't know are nurturing us. In an era where loneliness has both been deemed epidemic and also championed as the real key to self-actualization, *This Season, The Next* is a retrospective affirmation that 'There are a lot of things in life that are better when they're shared.' As much as that message resonates (not in small part due to a turkey you'll never forget), implicit for those in the midst of loneliness and loss is that this is a story that can only be told after the making that is a re-making, after risk, years and years after facts one cannot see clearly without distance from them, hope in Knott's Iowa a thing with feathers, yes, but bloodied ones."

—MATT MAUCH
author of *A Northern Spring*

"A serene meditation on a turbulent stretch of years, a soaring affirmation littered with dead birds, *This Season, The Next* renders memoir as a magnificent paradox. Blending meditative poetry and matter-of-fact reminiscence, Casey Knott doesn't just mend her broken American families, she reconfigures them as a hectic array in sunstruck stained glass. The illuminations are both a delight and something deeper, a prayer 'asking not forgiveness but space/ for mercy.'"

—JOHN DOMINI
author of *The Archeology of a Good Ragù*

"Casey Knott's *This Season the Next* is a luminous, lyrical memoir about the wild and wondrous margins of this life and what might rise from the ashes of loss and heartbreak until, in her words, 'the world blooms around you in ways you never expected.'"

—THOMAS MALTMAN
author of *The Land and Little Wolves*

"Casey Knott's *This Season, The Next* skillfully weaves growing a garden, raising chickens, blending a family, and appreciating the wonder of each moment in carefully crafted poetry and prose. Anyone--everyone--who's had to find their way in the world will relish this tale of seeking out and building a home, seed by seed, feather by feather, and tile by tile."

—GWEN HART
author of *Never Be the Same*

"In this memoir, Casey Knott writes about her life with a keen eye and an open heart. She writes about moving past difficult times—inch by inch—and making a new home after the jolting end of her first marriage. And she writes about the hard work and adventure of it all. Her story and her grace with language give us a lot to think about, a lot to remember."

—NICK HEALY
author of *It Takes You Over*

THIS SEASON, THE NEXT

A MEMOIR

CASEY KNOTT

CORNERSTONE PRESS
UNIVERSITY OF WISCONSIN-STEVENS POINT

Cornerstone Press, Stevens Point, Wisconsin 54481
Copyright © 2024 Casey Knott
www.uwsp.edu/cornerstone

Printed in the United States of America by
Point Print and Design Studio, Stevens Point, Wisconsin

Library of Congress Control Number: 2024935367
ISBN: 978-1-960329-49-3

This is a work of nonfiction. All of the events in this book are true to the best of
the author's memory. Some names and identifying features have been changed to
protect the identity of certain parties. The author in no way represents any company,
corporation, or brand, mentioned herein. The views expressed in this memoir are
solely those of the author.

Cornerstone Press titles are produced in courses and internships offered by the
Department of English at the University of Wisconsin–Stevens Point.

DIRECTOR & PUBLISHER
Dr. Ross K. Tangedal

EXECUTIVE EDITORS
Jeff Snowbarger, Freesia McKee

EDITORIAL DIRECTOR
Ellie Atkinson

SENIOR EDITORS
Brett Hill, Grace Dahl

PRESS STAFF
Logan Bidon, Paige Biever, Sam Bjork, Carolyn Czerwinski, Chloe Cieszynski,
Elyse Edens, Gwendolyn Goetter, Sophie McPherson, Kylie Newton, Natalie
Reiter, Mattie Ruona, Angelina Sherman, Holly White, Ava Willett

for Matt

ALSO BY CASEY KNOTT:

Ground Work

This Season, The Next

Coming Home

I always imagined coming home would mean a winding road
saddled by trees and greens, a driveway long
and thin. That's all I'd need.

Home would be what lies beyond that marking,
the way an arrow marks the eye. For years
I have seen miles of this story pass as strange. How the other half lives.

And now home, home lies up that hill beyond the curves and quiet.
The trees. A single lane cut from
a small kind of wild. So it was made. What it was—

my life had changed. Divorced and with a head that spun
on a dead-end road with two young hearts to feed.
I remember once a woodpecker fixed itself to my porch

all day and drummed. And too, all manner of feathers
marked my path like crumbs. The kids and I made bouquets
as my future knocked and dropped a line and I did not question

because it's always the hard times that gives a story legs.
This road, teeming with local squirrels and groundhogs, with
chicken hawks and the hoofbeats of deer known here,

it's a world that begets ours. That frames ours in a way
that coming home means making a deal with the wild.
To say this bounty is ours. Our hands not even touching the wheel.

TO GROW

The first year we planted raspberries, blueberries, black- and goji berries, asparagus, tomatoes, apple trees, a pear, pumpkins, a few varieties of gourds, onions, all manner of herbs, peppers, and squash. There was broccoli, spinach, and sprouts. Also, garlic and rhubarb, watermelon and cucumbers, various kales and radishes. Eggplant, okra, sweet potato, red potato, habaneros, jalapenos, corn. Cauli- flower, sunflower, daisies—more. All this to say we learned, in time, that less is indefinitely more.

Envision autumn weekends spent canning our vari- ous, inevitable goods in smooth glass jars that winked at us all winter from basement shelves. Meals, year-long, peppered—born from the crops we grew. I was looking to live sustainably, to tread light in my consumption of Earth's wares. I wanted to colonize my own lawn—all two acres of it—to teach my kids the value of working for something and in the process make them more conscious eaters because maybe a five year old that grew vegetables would appreciate them.

I'd recently sprung from a divorce and found Matt, this wonderfully patient, aloof fellow who'd postponed college for an education in the Alaskan backwoods—a former chef, massage therapist and coffee roaster, current salesperson and jack-of-all-trade—who was raising his daughter alone in town. I'll never forget the line that won me over; it was in our early hours of exchanging words, having connected on an online dating service, I'd asked what he missed most about Alaska, and he goes: *Hearing the wing beats of ravens slicing through the winter sky.* What better words to warm this poet's heart?

Here we are in the drunkenness of new love, purchasing a home just shy of two acres on the edge of town—our own creek, our own hillside, some woods, and everything possible.

We should never box our dreaming, as anything sought for or hoped for will materialize so long as we believe. But, what we want and what we need are often different things. I'm out here planting seeds, imaging future goats, a miniature donkey, chickens, and of course those basement shelves chock-full of canned goods. I'm blending a family—Matt with his 12-year-old daughter, Ella (Elle), and me with my four-year-old son, Fisher, and two-year-old daughter, Phoenix. I've been wondering who I was for the last 17 years. I'm all expansion. I'm a future of hanging flowers and herbs to dry. I'm planting oyster mushroom plugs, sprinkling the essence of morels on the east-sloping hills. Those mushrooms, with a mind of their own, never come. I find chicken of the woods and elf's cups, but never a thing my hands had placed. I forget to pluck the flowers or else they find their way to a family of deer's digestive tract. Hardly whatever I intend takes root.

TO BLEND

We should have seen the signs glittering like neon from a roadside motel. It took a good six months to close on the new place because the owner lived out of state and was inches from foreclosure. When that was settled and the lien on the property was taken off, the mortgage company told us they wouldn't provide a loan unless the owner installed flooring on the main level. No one had lived in the house for years. Prior to that, the owner's sons stayed there while attending college and pretty much partied it out. He agreed to have carpet installed and promised to return to town to finalize things and then he'd up and hermit, incommunicado for weeks at a time. This happened over and over.

No one understood why Matt and I continued to bother with it all—the house in disrepair and needing a complete remodel, the owner dragging out the process for months on end with excuse upon excuse so much so that when he claimed he missed a meeting because he had a mild stroke, not even his realtor believed him. But the run-down state of it meant it was something we could make ours, and more than the house, we loved the land it stood on—hidden almost on the edge of the city of Des Moines, and not in the suburbs, where I refused to reside. Matt refused the idea of moving his daughter out of her current Johnston school district, as she'd already moved plenty and dealt with enough in her 12 years. Lo and behold, this place was in her school district, and the address was in Des Moines, check and double check. We just had to be patient.

When it came to our closing date, and we arrived to sign the papers, we discovered another hitch—the owner's daughter wouldn't agree to resolving the lien that had been placed on the house until it could be proven her father had fully paid her late mother's alimony. We felt deflated, but sympathetic as our own difficulties with this guy were clearly part of his long-standing pattern. So, a new closing date was established to allow the courts to check that the late ex-wife had been compensated justly, and we were, at the least, given permission to access the house so we could start painting. Worrisome, sure, to invest on property that wasn't officially ours. But, divorced, and trying to sell the home my ex and I once shared, we needed to have room to settle into immediately. It was a risk I had to take.

I couldn't recount the last time I'd felt sure-footed. Nothing is guaranteed—at any moment things can change—a car might veer into our lane, or the heart might give—but still, we need to feel rooted in some way, to have a sense of our direction despite all of the unknowns. I needed a kind of closure—with the divorce; having been a stay-at-home mom, now with the worry of finding side hustles so that I could continue being home with the kids on the weekdays; connecting with Matt and blending our families; this move, and now this house. I needed to feel that the ground was stable.

So, we set about painting every inch of our fingers-crossed, future home. Weary and neck-strained from rolling a fresh coat on the ceiling, we noticed a tourist-like minivan in their ascent up the driveway. A young couple got out along with their four kids. They were having a look-about. Technically the house wasn't ours but still. We greeted them as surprise guests and discovered they were there at the behest of the owner, who had listed the house for sale on Craigslist and told them they were free to check it out.

I believe our gut-punched reaction scared them off, as they did not pursue their interest in touring the house even though they said they'd been in touch with the seller for some weeks. A few days later (half a year after our offer had been accepted), it officially became ours and I immediately saged the bejesus out of every lick of that space.

Strange, the things left behind in homes—baby teeth found in drawers, nudie magazines, and Polaroids circa 1970s stashed in the drop-down ceiling of the basement family room. Strange that you can live with somebody and pocket so many secrets, think you're the only one with a place to hide them—beer cans in the attic and throughout the garage ceiling: a nude calendar, a worn copy of the true-crime novel *The Menendez Murders*, a stash of notes—grocery and to-do lists with the phone number for the abuse hotline scribbled in the margins, and a marital counseling questionnaire with several responses written in bold, underlined: "HATE DIVORCE, START OVER, all he does is HATE, HATE, HATE." I won't speculate on who stashed what where as there has been more than one owner of this house and also, those aren't my stories to tell. But I will say my breath caught when I read those secreted words. I held space for the author of them, their message a kind of haunt I held in my hand and I sent light and love and honored them at every remodeling project—removing the HATE, HATE, HATE with "LOVE, LOVE, LOVE." Here Matt and I were adding to what we found left behind—the baggage of two divorcees starting over, histories clipped, a trailhead yet unnamed.

Other surprises we found—black mold behind the wall paneling, the kitchen plumbing dropping in my hand as I went to inspect a leak, the lower kitchen cabinets not even being attached to the wall and the countertops not affixed to them either. In fact, every faucet leaked, and worst of all, so did the gas line. On the phone with the gas company, we were told to exit the house immediately and stay put, that someone would be by in the next two hours. The fix came quickly, as the city takes gas leaks, thankfully, as an emergency. It was astonishing how little of this was noted in the inspection. The oven didn't work, the washer and dryer were a ticking timebomb; every wall, every ceiling had two layers of drywall so installing new light fixtures or switches required extra spacers just to get them aligned.

We had no business being plumbers, electricians, carpenters, tilers, and installers but watch enough YouTube clips, and you start to feel like an expert. My son, Fisher, was a huge help on our projects, even at a young age. Instead of watching superhero shows like a toddler, he was keen to turn on *This Old House* and the like, so his propensity for construction had been wired in him for some time. Even if he couldn't directly do the work, he was quick to grab the needed tools and assist, especially in the demolition side. He and Matt installed new bamboo flooring (after we tore out the new upstairs carpet, which had been placed in the kitchen, and which we'd be using downstairs) while I put together new kitchen cabinets.

The interior doors and trim didn't match the darker flooring we'd installed, so we replaced all the trim boards and I sanded and stained every door and frame to match, in lieu of purchasing new to save money and lower our environmental waste. We scored a gently used Viking gas range from the local Habitat for Humanity Re-Store. We cut our own kitchen countertops, installed all new lighting throughout the house, bricked, patched, and tore down walls, renaming them. Aside from paint and fixtures, the three bathroom remodels would wait.

At night, we'd sit in the mouth of the garage with worn muscles, with splintered fingers and bruised knees and look out at the dark lawn plotting our garden space, the fruit trees and berry patches to be laid, the landscape to work, and the animals to fetch. We were exhausted but we never tired, such is the adrenaline of having a dream within reach.

TO RENEW

Two things about our early days here stick out as having been quietly monumental. I should mention patience has never been a virtue I could dial in. When I'm hopeful for something, I don't like to wait. I do what I can as if my moving alone will make it come to fruition. Some random summer day I decided I'd had enough of the original musty, stained, 50-year-old carpeting downstairs, and I set about removing it myself as we were replacing it with the new fragments from upstairs. No YouTube tutorial needed. I made the kids popcorn. Three and five at the time, they were content to sit on the steps with their treat and watch the show of me.

I made the rookie mistake of pulling up the carpet from all sides of the room and piling it into one giant heap of a ball in the center so that not only was it too large to fit through the door, but much too heavy for one person to move. No problem, I sisyphused that boulder. I pulled, pushed, punched, rolled, and inch by inch, it *did* move. It took an hour to reach the door, where it became insufferably wedged in the frame. The kids, not knowing better, or maybe knowing more, rooted for me from their seats and with a lot of sweat and silent cursing, I managed. They erupted in cheers and claps and I turned to my daughter, cocked my chin and told her, *See Phoenix, girls can do anything.*

That night when Matt got home, saw what I'd done, and listened to my recounting of the strain it took, he said one thing: *You know it would have been easier if you'd taken a utility knife and cut it into smaller strips?* Which is exactly what we did then to get the damn thing off the patio. Still, that moment I locked eyes with my daughter, the look on her face that said, That's my mom—her having seen me move that boulder—made all of it worth it. It's synonymous with being a parent that we tell our kids how they can do anything if they set their mind to it, but there's a whole lot more substance to it when that sentiment is shown.

Another summer day we tore off the cheap faux wood paneling in Elle's bedroom hoping to find finished drywall, but we found black mold thriving there instead. We promptly ripped out the ruined drywall and patched in the new, but instead of taking the time to mud, tape, and sand it in place, I set about covering our quick fix in new faux wood paneling. She wanted her room back and who could blame her for not enjoying the wait. As if being in the midst of that awkward 12-year-old stage wasn't difficult enough, she'd already moved from Alaska to Minnesota and three places in Iowa, so she needed a space to plant herself, something she could make her own, a kind of longing for stable ground I knew well.

The paneling was large and awkward to carry and even harder to cut properly by myself, but I attempted anyway. When the kids were at their dad's or school and Matt was working, I needed to be useful and make the space they'd come home to more inviting, more homely—perhaps part of me was wrestling with my worth—another kind of baggage we carry after divorce. I stood bow-backed on the driveway with just a floppy handsaw and time, trying to make the small tricky corner cuts so that it would fit the puzzle of that wall, its switches and vent shaft. I just kept thinking, *I'm getting through, as long as I keep making forward movement eventually it will take.* I saw then how the universe kept bringing me face-to-face with what I needed to know. In this case, another lesson on patience and this time, I listened. It took a long time, it was a hard material to cut by hand, but I stayed put and kept at it till I had that thing just right. I really didn't mind the nickname Matt started using with me: Rough Cut.

There's a kind of faith in waiting. A belief that it will unfold. This notion has been pinging at my periphery for years. In my early 20s, years before I was ready to have kids, my daughter visited my dreaming and placed her name in the ink of my womb. In my dream, I was standing before the ocean, a barren beach before me and a girl who stood with her back to me. The name Phoenix rose up and out of my throat and I hollered it a handful of times amidst the noise of waves lapping the shore until the girl turned and looked at me, blue eyed. Mine. When I woke, I knew I'd have a daughter one day and I'd name her Phoenix. She is a kind of magic. She can sense the emotion in a room and find the words to better it. She is witty, determined, creative, and has an awareness of the world that is beyond her age.

THE FLOCK

The chickens came by happenstance. I'd found a Craigslist ad for a pre-made coop. For once, a break from building something after eight months of home repair. Bless the maker who sold it and their kid who painted a pastoral scene of some hen overlooking a serene field above the nesting box. All the materials that went to making it, we learned, were leftover building supply scraps from big box stores. So what if there were nails sticking out and gaps in the window frames—this thing had shutters and looked like a Lincoln Log cabin on steroids. Better yet, it came with four baby chicks they couldn't identify (two Barred Rock and two Wyandotte as it turned out), we named Juice, Jerry, Pickle, and Jaico.

If you want chickens that trail you around the yard and try to follow you in the house, that run up to the car when you pull in the drive, and are excited as a dog to see you, you must get them newly hatched and live with them. We kept ours in a dog crate in the office, a heating lamp on top for the first four weeks, and we showered those birds with attention. We'd cuddle them on our laps as we watched movies in the evening. We took them outside for picnics on a quilt every afternoon. Our sweet dog, Neko, who was rescued from a shelter, pre-kids, when I lived in Wisconsin, would sit near inhaling their scent, curious about every squawk and flutter. They learned to trust his presence.

Though we'd purchased the coop, they still needed open-air run for them, so we spent a long weekend attaching a 4x7 foot screened in space out of 2x4s and chicken wire that would allow them an outdoor space to roam safe from their many predators. We built a door large enough to be able to access the coop for easy clean up and water/food refills. At six weeks, with feathers fully formed, the chicks were large enough to move to the coop, and at that point, I couldn't take the smell of them in the house any longer. Even after cleaning it out daily, their barnyard stink didn't take long to go airborne, along with the down of their feathers and the dust from the pine shavings meant to absorb their frequent "droppings," all of which left a fine coating of ick in the room they were kept.

It's not that we consume many eggs, so our desire to raise chickens wasn't solely for protein, but rather it seemed the quintessential farmish addition. Chickens are lower maintenance compared to other livestock and their free-range eggs are much better than the store-bought variety—higher in vitamins and minerals, and lower in cholesterol. Another big draw was the fertilizer they'd provide for the gardens. Added to the compost pile along with a mix of leaves, straw cleaned on occasion from their coop, food scraps and yard clippings, their manure helps establish a rich blend of nitrogen, phosphorous, and potassium that garden soil craves. All it takes is sun, rain, and time to cure it.

Beyond the simple entertainment of having a flock, especially as they get older and their personalities show, we wanted to give the kids the experience of caring for them, be it gathering eggs, offering feed, or cleaning the coop, to provide them a sense of responsibility. Another nugget for their future nostalgia. It would also give them a lesson in food independence, as growing and raising our own food is more sustainable; the less food we consume from a store, the fewer greenhouse gas emissions via their transport we'd be supporting. And too, a huge plus, foraging chickens help reduce the mosquito population, quite pervasive around here when we moved in. Many evenings were too unbearable to be outdoors with the swaths of them relentlessly attacking. But chickens were said to scratch the grounds and eat small insect larvae, making for a kind of natural bug repellent and it most certainly does work. They eat the peskiest of insects, from mosquitos and fleas, to flies, ticks, and grubs, as well as the occasional mouse and supposedly small snakes—the latter I hoped for given the number I've crossed paths with in my yard. Unfortunately, ours do not have a craving for snakes. But for the insects, they surely do. The difference is evident when I'm at a neighbor's house—that constant state of swatting the insects away even after loading up with DEET. Within a year, the chickens had made our property noticeably better bug-wise. It's these kinds of benefits that make us willing to continue with chicken husbandry despite the heartaches and rough patches we'd come to know.

Note to Future Self

You'll come to the day split-knuckled
from the season,
a sky blushing like a secret crush
and all your thoughts, awash as the cloud
of turned up dust from rusted trucks
that pass too loud.
Much has changed.
Your old life cast off like snakeskin, where what remains
becomes warmth for birds,
treasure for some 8-year-old
boxed under bed amidst deer bones and odd colored stones.
Everything becomes an artifact.
And because there is a hidden part of the world
that spins some hearty magic—

watch where you step. You'll find these snakes
plainly awaiting your bare feet
like some holy ghosts, bidding
a new direction. Do not
disparage.
When the woodpecker offers itself to your window
someone new will not be far.
And when those jet trails form a perfect X
across the mark of dusk as you happen upon
your favorite stretch of grass in your yard,
the omen is a good one.
This life is stamped with memos, dispatched
from hidden places exactly when they should.

TO BLEND

We're in an interesting area here, a section of town known as the Dogpatch, named for the number of unwanted dogs owners abandoned here when it was considered the countryside before city development made it more of an edge. It's a tad lawless in that we don't have to abide by the same rules as those just a few blocks from here—we can burn leaves, have bonfires, and bow hunt deer in our own yards as most of the properties herein are wooded and sizable.

We have some characters. The first neighbor I met drove up our long horseshoe driveway one day when my son and I were just heading out the door to join my daughter, who'd stepped out moments before us. Imagine my surprise to see a large, beat-up SUV parked a foot away from where three-year-old Phoenix stood, and the first thing the guy says to me is this: *I wouldn't let your daughter out here alone. She could be eaten by a mountain lion. I've seen one in the area* (though not common in our state, every once in a while mountain lion sightings are recorded). Then he tells me he's seen a woodchuck in our ditch and that I should probably carry a shovel to chop off its head if I come across it, or maybe the man of the house, question mark, can do it. I wanted to say, *I'm more worried about you*, but simply thanked him for the heads up and squashed the conversation by replying in the most minimal of phrases till he went on.

Sure, we've woken up a few times to cars having smashed into the trees along our ditch, and we've found an exorbitant number of empty Black Velvet whiskey shooters at the edge of our woods, drive-by tossed, but trouble finds a home anywhere and most of our neighbors, especially those right close by, are warm and genuine and wonderful. We're lucky for them.

Around Halloween two years ago, the reclusive, junk-collecting neighbor, Rick, acquired a flashing roadside sign, the kind that blinks its warning—"left lane closed ahead, or construction for the next umpteen miles"—which he angled just so anyone driving on Lovington Drive might see it if they looked north of the road. School buses fence in the back of his house and a pontoon boat, 70s-esque, teeters near the front. The house itself with its macabre history as prior owners were busted in the 80s for locking their day-care kids in cages.

Rick programmed the sign to flash *Happy Haunting*, which surreptitiously one day stuck so that it perhaps will forever read this. For two years now I have appreciated driving by this fixed lick of words and thinking, 1. I need a picture and 2. of all the things to stick. It seemed fitting at first, delightful even, that this loner, who likely never had a trick-or-treater attempt the spellbind of his junk piles to reach his door for a bite of sweet, would offer that kind of nod to the season.

And then winter, the tailgate of spring, the charge of summer—that tome, *Happy Haunting*, became something else entirely. It became a point of reference, an inside joke, a notion. Sometimes it seemed a dare to be in my life more fully so that I was not just an actor delivering my part of a day's routine, but a narrator too, keen on purpose.

Another October would present itself and the meaning would change again—a kind of pinhole or an enlargement—becoming again what it intended. And now I heard from the neighbor between us that Rick died two months ago—a massive heart attack at work. He was found too late in the back sorting room of the local post office, now his sign seems like an *au revoir*, or a foreboding. The fact that two words could carry such weight, could change their meaning, and could resonate in turn with the seasons should rouse us. We cannot know the shape our words might take—how they land, come back to reckon. I never exchanged words with Rick, withdrawn as he was. The extent of our encounters was a quick wave if he drove by in his sea blue Astrovan while I happened to be in my front yard, but I appreciated his quirks, his flair, his sign fixed at the bend in the road. Now, crossing that sign on my way home I think 1. our words matter and 2. Rick, I hope you're not haunting these woods that lie between us.

TO GROW

The garden was established our first spring in the one ideal location on the property that wasn't all hillside or trees or creek. We didn't have the foresight for longevity and aesthetics at the time, as the pocketbook creates its own storyline, so we picked up a dozen four-inch by eight-foot wooden stake posts and some chicken wire to box in an area 25x30 feet, tall enough that the deer couldn't jump. I called a guy to come till the space, called another to haul in fertilized dirt, ordered asparagus crowns, and planted those along with tomatoes, green beans, cucumbers, onions, spinach, and broccoli. All the herbs (basil, thyme, parsley, oregano, mint) were planted in large pots on the deck so that I could easily snag them as needed while I cooked.

Probably the worst thing I ever planted was a goji berry bush that I buried a few feet from our front door, which never berried and shot up willy-nilly across the front of the house in no apparent pattern. I dug that ugly thing out a few years after nothing kept happening, but it still reckons there—little shoots somehow finagling their way through the garden fabric, rock, and mulch. Our asparagus sprouted and seeded and we anticipated loads of their stalks to be harvested in a few years' time as it's best to let them go uncut for the first couple of years. A wait, yes, but once established, they'd grow and feed us for 20 years at least before needing to plant more.

That pattern there—the repetition of years—that's how it goes, growing. You might strike it rich with a crop one year and not the next. You might wait four years for the apple tree to produce fruit, and when it does, the worms might beat you to it. So, you keep planning for the next and the next. There's a nuance to it, and it takes years to understand. Our tomatoes and beans were prolific, but not much else produced enough to be of meaning. Some powdery mildew overtook the broccoli, and our onions were the size of limes. We had plenty of herbs and lettuces, some cukes. And, oh my, the gourds we planted propagated the pumpkins we tried growing, but we had nothing but smallish green/yellow gourds, the vines of which overtook our entire fenced-in garden. Year one, noted. We used this knowledge to make the next year better.

That fall we planted a dozen or more raspberry plants—some at the edge of the backyard and some by the garden—as well as blueberry and blackberry. The property already had one apple tree but it hadn't been pruned or tended to for years before our arrival so it was three times the size of what an apple tree should be—hard to reach—and its fruit full of scab fungus. We planted four varieties of apple trees, along with one pear that we could nurture from the start, pruning them to keep them the ideal shape: short and stout. What I learned in tree planting was this: when the instructions say to dig a hole twice the size as the root ball, mixing in organic material along with original dirt, you should probably dig it even larger. We had maybe six inches dug around the planted roots, and it's taken more time than it should to start producing because our ground is clay-heavy, making it harder for tree roots to extend but a shallow depth underfoot, not unlike the way a stubborn mind stifles understanding. What I also learned was this: gardening is rich with metaphor.

THE FLOCK

Juice, Jerry, Jaico, and Pickle started laying eggs—the size of robin eggs at first, small and yolkless—our first real summer here when they were 16 weeks old. At this age, they were mature enough to free range, which meant we'd open the door of the coop late morning, and they'd wander through the woods, scratching for their meals, or taking interest in us whenever we stepped outdoors. Because we had held them so often, they were not the least bit skittish. They'd squat and let us pick them up as they free-ranged the yard and never seemed to mind when Neko dashed by chasing squirrels. They'd perch on the railing of our front porch and run for cover under the bushes that bordered it when hawks shaded by. Without fail, come the ink of dusk they waddled straight to the coop to roost in safety for the night.

We wanted to expand our flock as we had the coop space and property for plenty more, so there I am on Craigslist again looking for baby chicks. I found a listing that claimed to have a large variety of breeds. At this point, we'd discovered our birds were light brown egg layers, so we were looking to add Welsummer for their dark speck-led brown eggs and Americuana, as their eggs come in a range of blues, and as a plus, the guy was located in town.

It was straight up a scene out of a horror flick—Matt, Elle, and I show up at this split-level duplex rental on the southside of town, not a chicken coop in sight. A toddler-sized slide and big wheels paraded the driveway, toys strewn about but no kids in sight. The man opens the door, invites us in and I have never in my life smelled anything as horrible as that house. The pungent scent of feces, of urine, of *I-can't-believe-that's-even-a-smell* was haunting. I practiced my shallowest breathing as he led us down a dirt-crusted staircase, past barbie doll parts, baseballs, and flip-flops; where he had at least 50 baby chick like specimens in an assortment of fish tanks. He said he purchased the fertilized eggs online, hatched and sold them and I could tell by the shoddy, stanky space, where kids lived and played and had to breathe, that it was a desperate source of income.

I wanted to flee that joint immediately but at this point it seemed that purchasing his chickens was more like a rescue operation—to free these birds from that dump. We picked the three that looked the healthiest—two Welsummers and an Americauna, paid cash and beelined to the front door, gasping all the while till we got to the car. You know how the scent of a campfire soaks into your skin, your hair, your clothes? Imagine that tenfold. We named them Frenchie, Althea, and Joan Jett and burned the box they came in. They were quarantined in the dog kennel in the garage for two weeks. I didn't think to call in and report that guy to the city's animal control, too caught up with the demands of our home projects and too new in chicken husbandry, but someone else must have because before long I came across a headline revealing there had been 55 chickens seized from a home on the city's southside, most of them found in 75 gallon glass tanks.

I mostly love our chickens, but that group, having come from that kind of hell, having to be relegated to the garage to air out, to defunk and thereby having missed our kids' hands cocooning them on movie nights, the dog close by chancing a lick at their feathered vents, never lost their skittish nature. When we introduced them to our first flock, sure, they'd run for cover behind the border of my legs, but they never ran to me when they grew to size, never hopped into my lap and clamored to drink the cup of coffee I set beside me as I sat outside, never let the kids catch them easily as the original flock did. You have to live with chickens at first to imprint yourself on them so they have a kind of trust with you. The Barred Rock, Juice, was so tame I could put a tiny dog harness on her, which I did, and walk her by leash into my daughter's pre-school for show-and-tell. I could never have done that with our newer birds.

TO BLEND

There isn't a whole lot that scares me—I can break up a fight, hike pitch black woods, MacGyver an envelope to rehome a spider, cuddle the toad my son wants to keep as a pet and name Acorn. I just can't handle my reaction to a snake in the wild at all. I appreciate their place in the ecosystem, I do. The thing is I just never want to encounter them. Unfortunately, it happens all too often.

When my divorce was in process, I saw the change as an opportunity to become a better version of myself—to evolve and bloom. It was a coping mechanism in a way, that I could not only unstick myself from the life I knew, but flourish in the face of hardship and come out of it, eventually, not just stronger but more interesting, more in tune. Of the few friends that stayed in contact, and I'll say this—some of the friends you think you have that up and abandon you too at that time, bring a deeper kind of hurt than the divorce itself. To think you only mattered because of the other person you were with or that they only saw you for what you were when married and not what you could become—it's a kind of rejection that aches. I had one friend that'd been like a brother to me and his response when I tried contacting him to see about catching up as I was to be passing through his town, post-divorce, was that he couldn't see me because my divorce was "too hard for him." I never heard from him again, but I hear from my kids now and then how "dad's friend" had visited. So many breakings. You learn to move on quicker with each one.

One friend that still checked in on me those days invited me to attend a weekend crash-course to become reiki certified, which I did. I'd already noticed strange brewings by then. Everywhere I went I found feathers, be it finch or owl, like breadcrumbs that lead the way out. They were so abundant that it felt like a kind of call to arms, so abundant that the kids and I took to making bouquets of them, stashed in vases throughout the house. The universe telling me again and again—I would get through this; in time I would be on the other side. And then this—during that reiki class I saw a light flash in my mind's eye so white that it made me question if I had truly ever known that color. I saw my spirit guide dropping feathers for me to follow and when I did, I saw her motion to me as if to say I could guide my own sun to set.

I picked up a pack of medicine cards, which is akin to tarot but with spirit animals instead, and to this day whenever I spread them, I get snake as my armor. Snake: to shed their skin and be reborn—to be grounded in the vibrations of the earth: a kind of strength. I've had a snake attempt to follow me indoors, which I only discovered when the door I opened to pass through wouldn't close for the stretch of its vertebrae wedged there. I come across them everywhere here. One day I almost stepped on seven snakes as I walked down the driveway, to the garden, to the backyard. I've come across them at the mailbox. They've been in my herb box, sunning atop the bushes that fringe the front steps, curled up aside my patio, inches from my foot. It's become a long-running joke for my family—mom and the snakes. I'm the only one here with a phobia for them and the only one here that has so many encounters. I can't help my instinctual reaction, my instant scream, or the way I jump foot high and run. Apparently, it's quite comical. Even knowing what snakes represent for me, even knowing their importance in nature, I do wish I didn't have to be reminded so frequently.

The deer sightings are welcome. For years I'd been home with the kids, working part-time from home and managing well enough with Matt's income as our main bread source to afford us the time for me to do so. When Phoenix, the last of the crew, started kindergarten I was a mess for days; grieving the life I knew that would forever be altered. Who am I, what am I now? It's a wonder every mother shares when their kids start to outgrow them. It's not that I didn't look forward to focusing my energies on other things. There's just that sense that as a parent you have all of these firsts—first words, steps, and so on, but for everyone there's a last—last time you hold their hand in a store, last time you carry them, last one heading to school full time. And you need time to mourn and honor it.

I cried from the moment I left her classroom that morning to the moment I picked her up from school. But in between—the deer. We have deer pass through here all the time, scuffing our trees with their antlers during the rut, lazing in the tall grass along the creek, stealing from the bird feeder, bellying my flowers. The day I mourned the end of my stay-at-home era, I happened by the kitchen window and saw a doe and her two young fawns under the apple tree just beyond. I went to the porch and sat watching them, and occasionally they would watch me to deem that I intended no harm and together we spent a good half hour just being there, coexisting. It felt like a kind of offering—one of gentleness and love. I suppose one could claim these encounters are just coincidence and perhaps it's presumptuous of me to imbue such meaning into something so seemingly small, but I'd rather believe the world is full of unexpected offerings and connections meant to ground, impart, and answer us. I think it is the wiring of a poet—to draw meaning from the world, to observe life as if it is always answering the questions we didn't even know how to ask.

THE FLOCK

We have so many toads here that I often wonder if Toad Holler is a more fitting name than Dogpatch. On one count, Elle found 23 baby toads alive, wandering in our basement. They're so abundant come spring that you'll hear me exclaiming *Damn toad!* a dozen times a day because their sudden movement in the grass resembles a snake's until I realize—damn toad. My old dog, Neko, that sweet mutt of a mutt who loved nothing more than swimming or playing fetch, used to play with those toads. On rainy nights as Matt and I sat dreamy eyed in the garage, old Neko would shadow the toads, scratching his paw on the ground just beside them so that they'd hop and he would too, never taking his eyes off or harming them with his claws.

One morning I sat on the front stoop drinking a cup of coffee and reading the news when I sensed something was off and looked up to see where he was in the yard. Neko was there across the road in the neighbor's ditch and I noticed a car at the curve of the road hidden from his view. I couldn't do anything, helpless to the story that might be. If I hollered for him, he'd run home and be hit. If I didn't holler, he might stay put, might not. I sat there willing him not to notice me and come home, but he did. He stepped into the road just as the car rounded into view, hitting him full on his side. I was running to catch him in that knot of quiet. The driver got out and Neko jumped up, pure adrenaline, bounding straight for me. I'm patting my dog down, amazed at the lack of blood or misplaced bone, and telling this driver it's not their fault. I thanked them for even stopping to check and sent them off into their day. Miraculously, Neko didn't have any injuries as the car apparently hit him the best way it could, evenly along his side. He never ventured into the road again. Sweet Neko, who never harmed a chicken, who befriended toads and would sit motionless on the hill just so a passing robin might land and he might watch it. For all his gentle nature, he wasn't defensive enough to stop another dog from causing harm.

I'd gone to the store and ran into a high school friend I hadn't seen since. I invited him and his wife and twins to swing by, see what I'd made of things. When they came with their dog in tow I thought—this might not be good—but swallowed back the wonder when they said their dog had been around chickens before, even stayed on a farm for a weekend, and all was fine. I should know by now to go with my first instinct, the one that said *do not let your chickens around a dog you don't know well enough*. But we stood in the rain showing our gardens, our berries, the coop, our bit of oasis, while the dogs played, not a bit of interest in the chickens. I moved my unease aside and offered a tour indoors, the dogs still friendly with each other outdoors.

What I've noticed in the seven years we've kept back-yard chickens—all it takes is 15 minutes for tragedy to strike them, and it will. When the friends left after a quick tour, I noticed the quiet, noticed how Juice didn't greet us at the door. And then Fisher offered that he saw that dog from the window with one of the chickens dangling from its jaws. We set to searching and it didn't take long to find six dead chickens strewn about our acreage—poor Jerry, the one who followed me in the yard and would steal sips of my coffee or attempt to jump in the car with me whenever I opened the driver's side door, had succumbed in the chicken coop where she felt presumably was the safest place with the small opening, the size of a dog door. There were bodies in the woods, on the side of the house, and feathers wet and pasted in the rain in piles dotting the yard, marking the moments of their struggles. The only survivors were Frenchie and Jaico, who'd fled to the farthest corner of the garage, trembling under the work-bench behind cans of paint and wood scraps.

I let my friend know what happened to save any future chickens from his dog, presuming he'd be around some again. The real blame was on us. I should have listened to my gut. We were heartbroken with their loss, but it was an important lesson, one that'd be tested time and again—chickens have a litany of predators: dogs, of course, but also raccoons, foxes, lynxes, coyotes, owls, and hawks. Also, themselves. To lose some isn't a matter of if, but when.

Since we were down to just two chickens, who right-fully seemed terrified in the days after the massacre as they sat despondent on their perches inside the coop, not even venturing out for the fresh air of the fenced-in run, I decided the best course of action was to immediately add more to the flock. Safety in numbers and all. I found a Craigslist ad from a farmer who had dozens of full-size chickens for sale—thinking full-size birds could be added to the coop right away, experienced in egg laying and free-ranging already—and ventured out solo to hopefully give Frenchie and Jaico some peace of mind. I'm not sure what kind of thinking transpires in their bird brains, but I believed this measure would help them process their loss sooner, and with just two chickens too traumatized to lay eggs, our supply was quickly dwindling.

Our chickens had been laying an average of six eggs a week each, which we stored in cartons in the garage, only refrigerating eggs that were dirty and needed washing. Freshly laid eggs are naturally coated with an outer membrane that protects the porous shell from taking on bacteria. Washing these eggs removes that protection, and thus introduces the need for refrigeration, but unwashed, the eggs will keep for weeks at room temperature.

Most chicken folks claim you should quarantine any new birds and slowly introduce them to an existing flock so as to avoid spreading any disease, but we were in a hurry given the situation. When I brought home four farm-fresh full-sized chickens, we showed them to their new digs right away. Their retrieval had been like this: I drove 40 miles through open farmland and a few miles of gravel roads before I came to a ramshackle assortment of barns, a tiny cabin, and tromped through the rain-slogged field, just me and a round, old farmer, a dozen horses, dogs, pigs, ducks, and sheep, to the threadbare barn where the chickens were kept. I picked out the four that seemed easiest to snatch from the hundred crammed in there that ran amuck with nothing but piles of cracked corn and a bucket of water for sustenance, handed him cash and hauled them home in a dog crate, their stench and squawking spilling from the backseat. When I delivered them to the luxury of a roomier space loaded with mealworms, fortified layer feed, an array of seeds and grain and corn, they devoured the new offerings as if they'd never eaten anything but the wind that hollered through the boney boards of their old barn.

Jaico had always been a loner of a chicken and her crown so smudged flat that I always wondered if there was something scrambled about her brain. Even her clucking noise was different, more of a shriek. She didn't engage much with the others, and it's a true and heartbreaking fact that there is a hierarchy in every flock—to watch them establish their place in that hierarchy is to witness nature's cruelty. In this case, Frenchie made it known that she was at the top.

None of the new birds challenged her, but she still made chase, cornering them and pecking at their crowns as if to put them in their place. To intervene only prolongs the process and makes it worse. In the years we've had chickens, this has always been the hardest reality to watch, and they do it time and again—puffing their chests, spreading their feathers wide to appear large and looming, pecking, drawing blood, even mounting another to show their dominance—not unlike the show of power we humans experience.

In time, they learn their place in the flock and these displays of dominance cease. Frenchie became not just the alpha of the group but friendlier towards us too. She'd squat down in the yard when the kids would pass as a sign that she wanted to be caught and held in their arms. Sometimes at night when I'd peek into the coop for the nightly headcount before locking the door, she'd hop from her high perch to close the space between her and my hand. Life bellows on.

Pecking Order

I don't know why I'm so surprised
watching these chickens establish their place
in the order of things—and how true,
the ruffled feathers, the puff of breast, the prance,
the blood on those beaks—
this nature being nature.
For days I've introduced this flock in small turns
while the sky has been deciding itself—
sun and gloom, wet as a winter moon.

And now this moment bringing the new—
to vie for their link in this chain
means they must endure the shock of talons,
cornered and beaten down until
they submit, and even then it won't end—
every morning they will practice
the hierarchy of their bird bodies
in case their bird brains forget.
How can it be so hard to watch these chickens do
what we do?

TO RENEW

We made plenty of mistakes in our early home improvement projects, and still do. When it came to the basement remodel, we were so excited to rid the walls of the 60s faux wood wall paneling and open up the space by tearing down the walls separating the bar area (which in its early heyday must have been quite the swanky spot with its cigar vent, red shag carpet clear up the front of the bar, the built-in tender's stool—the kind of space that screams *key party*), we opted to leave our furniture, including a ping pong table and the years' worth of assorted kid toys where they all stood and just work around it.

I know now—always clear your work area, and two, sometimes you ought to ask for help. We did some things right, like putting up 2x4 support braces on the load-bearing wall, purchasing a solid steel post beam and one 12x20 solid wood beam to span the ceiling. We were so juiced up from our previously completed projects, the know-how and muscle strength that had to have given us, we each propped one edge of the 400-plus-pound beam on our shoulder and climbed a ladder to finagle that beam in place over our heads. We might as well have been wrestling a tow-truck. The thing wobbled in our first attempt and proved too much that I lost hold and watched it crash into the ping pong table, forever buckling its frame and requiring yards of duct tape to piece together the wood top.

The second time we attempted just about took off my head. Again, with the weight on our shoulders, arm hooked around its girth to keep it steady while we climbed our ladders and tried to hoist the behemoth, wedge it in. This time the wobbling was coming square at my face. My only option was to jump. I cleared a few feet before it bounced off my shoulder, landing just shy of my toe. It didn't do much physical damage; I was sore for some days where it hit, but nothing more than the soreness of any muscle post-workout. Yet, in my 40s now, I know how those injuries have a way of hibernating in our bodies like a 13-year cicada only to emerge someday down the road as a mysterious ailment. It's in my bones. It's very likely I'll wake up in the future unable to lift my arm.

Matt called in for some help from his work buddies—a handful of construction workers that knew a thing or two about lifting heavy things and I watched as it took them over an hour, veins bulging with profanities, to wedge that beam in place. And here I thought the two of us could do it lickety-split.

The rest of the downstairs re-do was blissfully straightforward, if not mundane. This was back when the top trend among do-it-yourself home renovation shows was to take a wall and cover it with pallet boards. The irony of removing that old paneling to put up pallet paneling wasn't lost on me, and pallets were so en vogue! They could be made into a dining table or a bench or a garden bed. One thing I discovered about using pallet boards—never bother with removing boards from a pallet. It takes splinters and hammers and drills and still those things don't wish to be pried. We attempted to sawzall the boards but that left too many unusable specimens. If you want to save hours of removing boards from pallets that you still have to sand, polyurethane, and hang, just get yourself to your city's pallet builder. They let you purchase the boards that they use to make the pallets in singles and are on the cheap.

I loaded the car and still spent a whole weekend cleaning and readying them to hang. I bleached them, sanded and coated them with protector, and nailed each in place. If not for the random assortment of kid toys, kid wrappers, kid-I-was-going-to-use-it-still glass and plate strewn about the basement, the wall looks great.

TO BLEND

I almost didn't meet Matt. Having lived away from the area for a decade, with my few close friends in other states and the nearby few being busy in their own lives, the evenings when my kids were with their dad were especially lonely. I read a lot, wrote poetry and songs. I picked up an accordion at a pawn shop and taught myself how to play. Having never played an instrument before and no understanding of keys, this meant I tinkered with the sounds I could make and then I'd sing along, memorizing the keys and buttons I pushed. Still, I craved conversation and companionship, and needed to get out of the house to check out a show or dinner spot now and then. There are a lot of things in life that are better when they're shared. I wasn't looking for love but I was looking to expand my social circle, so those childless nights weren't so quiet.

My neighbor suggested an online dating site and so I spent one Saturday night creating my profile, filling out those silly surface questions about my interests, my intent. I'm wordy and didn't quite finish. Still, when I logged in the next morning to complete the final questions, I was surprised to have messages from potential matches already, and two, a list of profiles recommended, one of them being Matt. I sent him a note, something like *hey, I see you're a Vikings fan. I recently moved from Packer country— maybe we could watch a game sometime.* I didn't even care about football so I'm not sure why I led with this. And because I'm wordy, I realized I had more to say and sent him another message, something about all the ridiculous questions they make you fill out on this site.

Now I hadn't dated for several years but I understood there are unwritten rules for these things, and I knew sending yet another message might be a faux pas. But I figured if he took issue with it, then he wasn't right for me anyway. I was in my 30s, a mom, opinionated, not even looking for love, so send again I did. He replied shortly after saying he was out and about but would respond more later. That night we messaged one another for three straight hours, talking about kids, the city, music, food, and travels. The next night was the same—hours of exchanging words, and when I asked him how he handled responding to all the people making contact (some real creeps too—at this point I had over 100 guys messaging me) I did so mostly to light a fire under him as I was too eager to wait long. Our first date was the very next night.

Gravity was strange then, like I was static cling. The waiter spilled a tray of drinks behind me, Matt kept dropping his knife and fork as he ate, nervous in words. As we walked to a bridge after dinner, autumn stars and not a lick of wind, a lone strip of breeze blew swiftly and suddenly along the sidewalk. I could see its passenger from two blocks away—a single feather, which winded toward me and stopped with the breeze right at my feet. Some things are just known.

Two weeks later he would go in for surgery to remove a fistula in his gut. He almost didn't respond to my litany of messages that first day of contact because he couldn't imagine anyone wanting to stick around for the surgery and the healing it'd take. In fact, his Match account was set to expire and he wasn't looking to renew it any time soon. When he told me that he had Crohn's disease, that most of his colon had been removed and he would forever have an ostomy bag instead of a working digestive system, that this surgery might knock him down for some time, I truly didn't think twice about it. Everybody has something they carry. I was there when he came out of recovery, and I've been there ever since. We found love when neither of us were looking because the universe is funny like that. He'd tried meeting "the one" via dating sites for seven years and had given up due to health reasons and the resignation that it just wasn't going to happen, yet here I come along after my one day of being on that site just looking to meet an interesting someone to hang out with.

We were married our second summer here. I'm not one for fancy, for a big to-do and the money it takes, and at our age and with kids and this urban farm we were nurturing, we wanted to keep costs low. We rented a lovely stone shelter at a park in town, asked a friend to officiate, another to play music, and one with a passion for photography to shoot ours. We made our own food, bought a cake, borrowed tablecloths and decorations from a neighbor, and rented chairs. That was pretty much it. I purchased an assortment of flowers from Trader Joe's to make my own bouquet, and the morning of, we went to the site to decorate and haul in food and chairs, hang lanterns and feather garlands with the help of our parents and a few friends. We kept it small with just immediate family and closest friends. Phoenix and Elle were my bridesmaids and Fisher was Matt's best man. I'll never forget the five of us after the ceremony was complete and our vows cast, how we huddled there at the fringe of seats before receiving our guests, every one of us sobbing and hugging, the bubble of pure joy hitched in our throats so that we couldn't find the words to speak. Complete elation.

Blending a family certainly has challenges, as all families do. There are growing pains and lessons you can't figure out until you face them, but we all truly love one another, and with that at the center, we always come through.

THE FLOCK

For all the joys of gathering and cooking fresh free-ranged eggs, their quality far surpassing any egg found at the grocery store, for their knack at reducing the bug population, a welcome gift come Iowa summers, and the thrill of having those feathered friends run to greet you when you step outdoors, the chosen one. It's not all rainbows when you have a backyard flock. Chickens are loud, stinky, and they crap all over the danged place. No little finch excrement you might find dried, a small blotch on your windshield, these babies are the size of golf balls, barn-scented, and sticky wet. When you like to walk barefoot in the grass, their frequent dollops are hard to miss. It seems quaint almost when the chickens take to loitering on the patio, ducking under your legs when the screech of a hawk cracks the sky, but then comes the daily ritual of hosing off the waste they leave behind.

Forget about landscaping the front of the house with ornamental plants, flowers, and mulch. Despite the woods around here, shaded and plated with all manner of bugs, our flock seems to prefer to scratch for their food in our mulch. Their foraging is incessant—they'll use their claws to scratch the ground, flicking the mulch several feet behind them like a tire kicking up dust on a gravel road. Back and forth, one leg to the next, they scratch, scratch, and peck the ground, eyeing some microscopic treat. Having mulch means spending every evening after they've gone into roost with rake in hand, tidying the long stretch of grass they sent that mulch flying into.

When you wizen up and cover the area with rocks instead, that's when they head to the neighbors. Mary had been here long before us and is the kindest, no nonsense, shoot from the hip sort of woman that won't just lend you a ladder but offer to hold it steady. When Mary complains about the dang chickens decimating the manicure of her mulch work again, the kind of work she savors for her propensity to be outdoors, Mary gets a fence strung up between us. At first we tried other methods—cinnamon and cayenne sprinkled atop as research claimed it was a combination chickens would avoid. Not only did that method *not* work, but it also made her cats sick. So, over a weekend we fashioned a fence between our properties using chicken wire and whatever posts we had on hand.

During my stint in Green Bay, an old neighbor of mine, Randy, relayed the story of how the fence dividing our homes there was born. The old tenant of my home hated Randy's dog so much that he took to poisoning it—caught red-handed sprinkling rat poison in the dog's outdoor water dish. When Randy took to installing that picket fence, Mr. Dog Hater stood by repeating the line from Robert Frost's poem, "Mending Wall," that goes: "good fences make good neighbors." Whatever its history, I was thankful for that divide between us as his wife, Faith, liked to spend every weekend once the weather hit 50 degrees out on their patio wearing tube tops to catch a tan, drinking her Michelob Ultra from a crystal flute, cocking her BB gun at squirrels who dared to enter her garden whilst gospel music blared from her radio.

But this fence—Mary's—airy with thin wire so we can still nod across the way, commiserate over kids, politics, and the state of the world, doesn't feel like a divide. It does, however, keep the chickens out of her yard. If you're going to have chickens, be mindful of the havoc they can deliver to the neighbors—crack of the morning rooster calls, pillaging gardens, dropping their waste, and plowing through their landscaping work. Give them eggs. Keep the chickens at bay.

The chickens don't just test the boundary between our property line and the neighbors', but the windy road that hems us in. There are quite a few reasons why a chicken will cross a road, food being high on the list of their dare. The first flock, pre-dog slaying, chanced the blind curve of the road out front to ravage the neighbor's ditch-wild Brown-eyed-Susans so often that I dug a few of them up by the root and planted them along the backyard fence in an attempt to keep them interested here. Never mind that *here* already meant their own ditch-flowers, and woods, shrubs, compost, scratch, and organic fortified layer feed. Those birds never bothered with the Brown-eyed replants, but I suppose enough of our running to the road to shoe them home and the enticement of breadcrumbs we tossed to lure them back must have eventually planted some Darwinian awareness because they stopped their wide-roaming ventures not long after they started, but not before I caught another cross-the-road temptation.

Who knows how many days it came to sit there—sometimes you get the sense that there's a lot of plotting and waiting in the world of chicken predators even though experience says their attacks happen via a one-sided kind of luck. No one can tell me that raccoon camouflaged on the unkempt wild of the hill across the road was not channeling the Pied Piper. It sat there half-hidden among the wild shoots watching the chickens who meandered near the road so motionless and intently that its watching seemed a kind of trance, bidding they come. So fixed were the raccoon's eyes on the flock, it didn't notice me noticing it. Coaxed back to the coop and locked safe for some days, the chickens lived for then and the raccoon gave up its post. It seemed so encompassing then, all this chicken in the road business, but one minute we were living it and the next we were remembering "that time." From this end it was nothing but a short-lived season.

In other ways, to have chickens is to add another hitch in our own free ranging. When we have set plans that we know will keep us out for the evening we're mindful to not open their coop and let them have the run of the lawn, but sometimes the kids let them out anyway and sometimes spur of the moment opportunities crop up and you discover soon the chickens make you perpetually late for things. There is no reasoning with a chicken that they need to get roosting early for their own safety, which means loads of time spent searching their whereabouts, coaxing them to follow with the extra allure of watermelon slices or hunks of bread, a trail to their coop door, means chasing after them as they catch on and hightail it back to the bramble on the steepest hillside. It's impossible to catch a chicken that doesn't want to be caught—not so for a coyote or fox, but undeniably for us.

After a while you give up. Offer some curse words and resignation. You think to outsmart their predators by placing an AM radio with the volume set to max near the coop in the hopes the sound will keep those kinds of critters away and then spend the evening away wondering what survives. Later there will be flashlight headcounts, maybe some more cursing, the door latched tight.

TO GROW

Every year I try to grow pumpkins and all I ever have to show for it is one or two okay ones come October. The vines are massive, spindling out in all directions. No matter where I plant their seeds, their shoots will reach the garden fence, will climb right over, will claim the other beds. Squirrels are the hardest on them—every morning new claw marks, new bite marks; so that by the end of the season only a few plump fruits remain. It's a lot of real-estate they acquire, but I can't help but try as I'm a sucker for autumn—the dried spice of leaves, the warm bursts of color, the apple picking, the cooling of stars, woodsmoke from a campfire.

We close the garden too—the roots pulling way, dry as a bone, the stalks and stems tossed into our compost heap to break down, become better soil. We add our old chicken bedding there, cornsilk, and leaves. We've learned our plants do best in raised beds filled with composted soil, so we've added patches of raised beds throughout the garden, using mostly whatever scraps of wood we have that are long enough lying around to make the boxes. Even then, not everything grows—we'd given up on broccoli, cauliflower, cabbage, onion, garlic, and beets because they never grew to the right size and the thing is, the space they took or the time it took to seed and weed and water all for a few handfuls of skimpy specimens just wasn't worth it. I could go to the farmer's market or to any grocery store and buy a bag of these items for less than it took to grow them.

There's a kind of refining you realize gardening—that is to see what's worth it. The older I get the more I see how true that is for life. I used to think an abundance of friends, of things, or trips and activities was better. That there was a kind of success to be found in that overflowing, but those schedules fill up the calendar and it's hard to sustain any one thing or give it due diligence when you're pulled at the seams. Nope. Less is more. To live by that is to discover and have space to nurture your strengths, to grow best what you can.

There are plenty of people out there who can grow those things I can't and who excel in ways I could never fathom, never even want to. So why not be our best version and share in the rest, the lot of us filling in for the spaces we lack? I can grow all manner of berries, tomatoes, potatoes, cucumbers, squash, and beans. Maybe someday, pumpkins. Besides, what good story isn't sapped with hitches, with conflict? Never more true than the quandaries of a vacation.

One of our first vacations as a family—some rental that a late-night internet search dug up under the parameters that it was cheap given our dwindling budget from all the home projects—was at a cabin high in the Colorado Rockies. Despite the coin purse, and because of it too, you need to catch a break now and then. GPS led to gravel roads, windy and narrow, passing wild horses and middle-of-nowhere ranches. The cabin itself only accessible the half of the year it wasn't swamped with snow. The lone driveway, a blink and you miss it, from the main road took a half hour to navigate—single lane, twisty, muddy from rain, and the steepest grade I've ever ridden. The main gate that portioned off the rest of the stretch to the cabin itself was locked and the code provided wouldn't take. After so many failed attempts to open it in the wind and rain, I climbed the gate to have a look about the cabin and see if I could at least open that door.

With Matt wracked and white-knuckled from driving, and the kids anxious to roam, I left them waiting, cussing the weather, the stupid gate, and eventually the front door that took another hundred tries to open as the wear of its climate made for one jammed and rusty lock. I rattled with the frustration of it all here when I just wanted to plant some quaint memories for the kids, to catch a break in this mountainous splendor only to have it stalled. We got in. We spent that week in the boonies with no running water or electricity, just a small derelict shack with two bunkbeds, a wood stove that Matt overfed, and an outhouse, nudged in by a stream and surrounded by woods and wildflowers and bear scat. Stars, visible for once with not a single streetlight to obscure the view. The simplicity of a day spent waking and walking and wandering around. I spent the drive back home laughing at how ridiculous I must have seemed—rain-soaked and scrambling for almost an hour to unlock the week that of course eventually happened.

Every "remember that time" is sprung from adversity or surprise. It's a wonder anything easy comes. We take yearly trips in cushier spots and with far fewer hitches but that one is the most memorable. Phoenix in the outhouse claiming *I know why they put the bathroom here, that view*, making the best of things in a way only innocence can.

A Refining

I would have thought there'd be more—
bloom and fruiting,
heirloom varietals pinned
to the bed, feeding and dying
in turn in all customs of sky—
undulating as a universe,

but every season
is a harbinger of failures—
whatever the deer can't reach
the beetles and aphids flourish in.
The nitrates are off,
the pH all wrong and I'll spend long
hours exhuming weeds
all to harvest a few handfuls
of things. 3 feet of space
for 3 ears of wormy corn
and the pumpkin vines took out
a fence with the weight of its lone
fruit—orphaned by thoughtless
squirrels.

So every year I plant less
and less again—
a culling of sorts,
to see what's worth it.
Like you, I'm learning what takes root
is the thinnest thing.
Even the clouds that spill across this bit
of blue have a mind of their own,
as if whatever you heard about art
imitating life is wrong.

To grow is to retreat,
as home is the sharp tip
of an arrow welcoming its arc.
The story ends and you, no worse
for wear.

When you get to this breech
it's with a tuned hand narrowing,
starved blind as first love,
and this practice of failing—
what can you do but grow
the best one thing you can?

TO RENEW

We built a treehouse for Fisher on a slope of hill just along
the backyard where the woods started shading. We rarely
have a plan and never a blueprint for our builds, instead
we tend to discover what works best as we go. It was a
simple two-tier platform with fenced sides and no roof
that he, Matt, and my dad strung up over a weekend. In
time, Fisher would add salvaged wood scraps to enclose
the sides, leaving little pockets of openings where the
light might shine in or his peeping eyes might look out,
as well as several hundred dings from his BB gun, the
perfect target.

Our property had a bridge that once crossed the creek
to connect to a smaller flat bit of land on the west side,
but the bridge had long since caved in. And we crossed
that creek often. We donned mud boots and explored all
up and down its banks, discovering minnows, snapping
turtles, antique bottles washed up after storms, and always,
frogs and toads, and unfortunately, sometimes snakes. The
kids built lean-to's and a teepee out of downed limbs on
that flat tract of our land west of the creek. They spent
hours pretending they were survivalists in the wilderness.
A bike trail access was just beyond at the tail end of our
neighbor's property line, and we accessed it frequently to
ride or hike to the woods and ponds they lead to some
quarter mile away.

We needed a bridge and had all the materials at the ready—some felled trees, two trunks equal in size, and the boards from the fallen bridge, which we pried off to reuse. The problem was moving the trees to span across the creek. At 30 feet in length, hundreds of pounds, and no way to get a truck or tractor close enough to help pull, we discovered the only option was ancient—prop the ends up enough that a log might fit beneath and roll. One can lift a bit of tree but not the whole. It's a slow process placing log rollers, pushing the tree forth, and repeating and repeating. But it worked, and it was just Fisher, age eight, and I who moved them to the bank, waiting then for Matt to get home so we could manage the final crossing. He and Fisher worked till dark and then with headlamps, worked into the night nailing in the plank boards, making us a bridge.

A few years later heavy rainstorms would deliver six inches of rain, causing flash floods, slogging the creek, and washing the bridge out before dropping it 50 feet away, cockeyed on the flat land west of the creek. A small wine bottle would rest at the center plank, like a note left in a bottle, tossed at sea. This one would have read: *The only certain thing is what your heart alone brings.* For its time, that bridge was a lovely thing to have. We still haven't found the time to rebuild it as days, years, ebb and flow, reshuffling our priorities.

There was an old cement slab at the edge of our backyard where we envisioned a shed. There was no way to have a pre-made one delivered to that spot, so the only option was to build one. YouTube, homemade trusses with the help of my engineer brother-in-law, and a few long weekends. We had no blueprint but erected it organically according to the space—profound in its build, this shed. I learned this in the process—the frames for walls are wobbly and can easily topple over, but attach the roof trusses and those walls are solid things. It may have been my mind being unhinged with all the projects, planting and parenting, but I was finding metaphors in everything. This one—the thoughts we choose are the body's framework.

TO BLEND

Blending families to be a cohesive space where every party feels equally valued and loved is not a given scenario for all. My good friend was raised thinking all stepdads and stepsisters labeled their food items in the refrigerator, not unlike an office space, to clearly mark each thing that she was not allowed to have. In fact, her stepfather explicitly told her that he would never be a father to her. Her childhood was akin to a summer camp that never felt fully home—more so a temporary arrangement. Years would pass, and that same man would eventually walk her down the aisle on her wedding day, but that kind of bond took a whole lot of time.

Thanks to her, and my own ex who grew up with divorced parents who could scarcely share the same space and his gumption to do things differently, our kids adapted in a way that only kids can—wholeheartedly. We made sure our kids never feel guilty for choosing which parent they went to for a slap on the back job well done after a certain ball game or soccer match or the litany of school events by sitting together. Every child wants a home and how they find it is their story, but for our part we did the best we could with what we knew. The most important thing we did was love them without measure and to recognize our own imperfections and deal with them accordingly when they sprang.

For one, it seemed my post-divorce sense of not being wanted, met with a 12-year-old girl, who had little trust in women and a whole lot of protection for her father whose health was questionable made me feel like his companionship was a kind of battle to win. I wanted Elle to feel like a kid and not an adult partner traversing the highs and lows of life, worried about the future. But this is all she'd known and here I was nudging her to let go and be a kid. Wrong of me to assume she'd want to pass the torch. The best I could do was not press. Let her be who she is and let her see me equally give time and attention to all the kids and to Matt. All three kids went to therapy when they showed hints of needing some other attention they weren't getting from us—a quiet brewing—and I know enough to know I can't know everything. In time, the edges began to fade, which had mostly everything to do with them—not Matt, nor I, nor our exes.

The next hurdle is getting the kids themselves to gel and that comes in time too. One of our biggest obstacles in that was getting Elle to respond as a peer and not another adult in the house telling the younger kids how to mind their business. Even something as simple as a family dinner can become a statement of will. I'd press on Phoenix to eat a few more bites of food only to have Elle usurp my authority and tell her if she didn't eat another spoonful she'd be sent to her room for the remainder of the evening. She'd present these ultimatums to the younger kids often enough that I worried they wouldn't see her as a confidant—years later swapping stories of their youth-sprung shenanigans and truths to bind them—but instead another adult to duck and avoid when they wanted a taste of freedom.

That she could be the big sister Fisher and Phoenix might confide in, the one to answer the questions their young minds wouldn't ask a parent, so they would grow to see Elle as not just a safe space for their wonders, but a role model, complete with her intelligence and self-respect, her care for school and kindness to friends—at a certain point it stuck. Those three now have that assured nod they give each other, the kind that says, *I've got your back* to whatever bully life throws. Like the garden, these connections took time, took nurturing, took roots.

Sure, we parents had a hand in giving them this space to grow in, but what it really was is these kids made themselves at home. For Matt's part, he always wanted more kids and now he has them. His patience, his wisdom, and his dad jokes have made it so Fisher and Phoenix see him as another great father figure that they look to for his healing touch and know-how. Matt has taught them to throw a football, to tinker, to show respect. Only the future will decide what else.

It's not to say we still don't have tears in the stitches we've sewn together and honestly there are times Matt and I grow defensive of the children we brought to this home. When you're the bonus parent it can be hard to reprimand the kids' behavior—you have to do it, sure, lest the kids start to take advantage, but you also have the weight of not wanting to seem like the real-life version of some Disney-esque evil stepparent as well as the potential for offending the parent who provided the DNA. When you don't start out as the co-parents with your methods and inclinations for rearing your child established from the start, and no shared history of the child's early experience, it's more of a challenge to align that down the road because how do you shake that sense that you "know better," having been there all along?

It's been easiest with Phoenix as she was so much younger when we all met—more adaptable because of it, still so many firsts ahead. But we're all works in progress. The more the kids see the adults in their life always working to become the best version of themselves, the more respect they'll have for that adult. Emulating the traits we intend to instill—compassion, grit, self-love—is the best we can do.

THE FLOCK

Matt had been hinting at getting another dog. As a mom, three kids, a brood of chickens, and always a thing to do around here I held firmly against it, but with Neko getting up there in age and clearly not cut out for protecting chickens, I caved. Then I did apparently what I always do—search Craigslist for another animal. I found a post for Anatolian Pyrenees (Great Pyrenees and Anatolian Shepard) mix puppies and drove solo across Iowa farmland one afternoon to pick ours. These were the directions I was given: when I got to the named rural highway look for the farm on the right side of the road, white house, no electrical lines, ask for Yoder. So the ad lister was handling the post for their Amish neighbor who obviously couldn't be reached by phone or computer.

When I arrived, Yoder greeted me at the door, hollered for his dog, Cutie, and took one look at my shoes to see if they would handle the walk to the pups. The three of us slogged through the muddy yard, crossed a corn-harvested field, and up a steep hill where a rusty teepee-like drum was placed, open to a wide sky, no trees for windbreak or shade, a baby pig carcass plopped in the muddy entrance. He asked if I was interested in a male or female and proceeded to grasp three of the latter out by their nape as a mom might do, and one of them walked straight to me, sitting down on my boot. She was it. I scooped her in my arms, paid Yoder, and was headed home all within 20 minutes of arriving. You can bet our new puppy, who we named Luna, had worms.

Strangely enough, I had just finished reading *Where the Red Fern Grows* with Fisher. We had passed the last page of the book back and forth to get the other to read as neither of us could finish for the crying it invoked when Matt ran in exclaiming how Luna had treed a coon and wouldn't budge. We aren't hunters but we have a pellet gun for target shooting and chicken safety. We've seen raccoons trying to breach their coop on our trail cameras, and we've seen loads of their paw prints surrounding that joint and all along the creek. We knew how stubborn Great Pyrenees were, knew Luna would sit through the night and the next day at the base of that tree, barking and waiting, so Matt shot at the raccoon and knocked it square in its hump, the jolt of which brought it scrambling down the tree where Luna sat.

Great Pyrenees are an ancient breed and their 2,000-year-old bones have been discovered across their namesake mountains in France. They were bred to live alone on those mountainsides and protect goats and sheep from wolves. In the modern world, they still do, but also as guardians for livestock, chickens, farms, and families. Whatever isn't part of that is fair game, so that raccoon didn't stand a chance here. She knew the best way to kill—go in for the neck bite, fashion her jaws clamp-like, and shake. She came away without a scratch or a drop of blood on her, and we came out with our shovels to bury the first in what would become the coon graveyard.

Pyrenees aren't considered to be fully grown until they reach their second year, so being a puppy still, we always brought her indoors when we left the house despite the chickens free-ranging in the yard. My daughter had a piano recital at a nursing home not far away, which would hardly take long, and besides, it was midafternoon on a Saturday with plenty of neighbors out dawdling in their yards.

Yes, I'm keeping up with running the kids to and from school, to playdates, to Girl Scouts and Boy Scouts, baseball and soccer, which I coached, guitar lessons, piano lessons, swim lessons, all the while performing the perfunctory tasks of domesticity—bills, cooking, cleaning, laundry, shopping. I'm working part-time scoring essays for an education assessment company, I'm a poetry editor for a literary magazine, and working part-time in the school, and in all of that looking for a bead of time to write, which, aside from being a mom, is all I want to do.

Not an hour had passed before we returned from that piano recital and when we did, we caught the last breath, the last flutter of one of our chickens dying in the front yard. We let Luna out to help us locate those that might have remained, and she did—running the lot, sniffing frantically at the ground, discovering the remains of four more. Matt saw the fox we believe did it—he'd walked the trail that led to the bike path beyond and looked square on at a fox that seemed to lure him to one side of the trail, smartly keeping him from looking left, where its den was hidden, where no doubt a litter of mouths nipped and fed on the one chicken we couldn't find. It's hard to feel bad at nature doing its nature thing—where one loses another is sustained. Such is life. Of the few that remained, was Frenchie. We Craigslisted, this time driving together, Matt and I, to a farm in Jamaica, IA to expand our flock once again.

Despite All My Effort,
I Am Nowhere I Thought I'd Be

Here, at this intersection awaiting
my turn to pull ahead—it's half-light now
and we are supposed to decide this part.
Already some have stopped home

and now they're walking their dog
or else pulling weeds from their beds
as if to say they own some thing.
Dinner is on the stove and their kids

are spying on the neighbors from upstairs
rooms, blue-lit and sprawling with their own
resentment. We have no idea
what they think of us.

1   2   3
I'm breaking. I'm inching forward,
as if it'll take some wild bird to alight
my knuckles and ready the way.
Divining rods. A checkered flag assailing

from the blues of you.
The car ahead moves much too slow.
And these monuments for living—
the road, the kids, the grace, the aging,

it's a snare in our ribs
fixed as a star, which we wait all day to
see in the dark veneer of nightfall.
And yet, the more

things change, the less we change—
as if to stake some ground in this
wheelessness.
A route our own.

I'm at the corner now.
Your dog with her holographic eyes
is barking at the squirrel with the entire
husk of her body. The squirrel is retreating

with the entire husk of hers.
North is north of here.
A compass spills from my heart and it
can't decide which way to head,

but always, it knows
the direction home.

TO GROW

Winters, here, are my favorite. No poison ivy, no buck-thorn, or ticks, or snakes. I can walk freely in any part of our property without wonder. The chickens stay inside the coop and run area, which we cover with tarps to block the snow and wind. They're keen to stay cozied in, laying fewer eggs because of the cold, and probably practicing their hierarchy still. We give them plenty of straw and keep their water from freezing with an electric heater.

With the right snow gear, you can stay put outside for hours. You'd think by now we'd have a snowblower at least, or a blade for the mower as the half-moon of our driveway is 200 feet in length, but we like to shovel. My joints get achy sitting too long indoors and I like the exercise of scooping a shovel, sweating, and cursing at the busted-up bits of our driveway that the shovel always seems to hit, the ghost of my breath following. I like feeling exhausted and accomplished when we're done, some days making the pass after every few inches of snow even while it's still whistling upon us because a few inches a handful of times is easier to manage than the weight of a foot. Only once did we flag down a distant neighbor as he passed in his bombed-out truck with a blade to ask him to plow our driveway—15 inches overnight and wet and heavy. When I asked what he wanted for payment afterward, he smiled, toothless, with whiskey-shaking hands at the wheel, and said no payment needed—*That's what neighbors do.*

We start campfires on the colder days, sled our front hill, and retreat to the back fire, cooking hot cocoa over its flames. Skate-walk the creek, hide and seek in the woods. When the kids go to bed, Matt and I like to sit on my favorite hill and just be with the sky tinged light orange from snow and streetlights. I've made snow chairs, wine glass holders there, watched the moon rise, written lines for poems in my head that I hoped to remember. Once, while I was away visiting my best girlfriends for a weekend, Matt and the kids found a raccoon frozen in our wonderland of woods, its heart ripped out, clutched in its own hands. The only thing you can do with a dead raccoon in the winter when the ground isn't diggable is place it in the garbage can and hope the temps stay well below freezing before garbage pickup day so that the stench of its rot doesn't reach.

Come New Year's Eve, we toss the dried-up Christmas tree and empty boxes on our burn pile, and a tradition I started our first winter here, we write our wishes on scraps of paper and set it all ablaze. Then the five of us stand back and watch the flames turn our dreams to ash to be carried on wind, becoming atmosphere, believing that the universe will answer them back into our lives when it sees fit. Even now, years in the making and Elle off at college, she'll ask when we're burning the tree and come home to take part in our ritual. It's one thing that has stuck.

The mark of our steps through the snow from the house to the coop hardens with wear so that it becomes a trail of ice, pockmarked and treacherous. That bone-breaker trail will be the last thing to melt come spring. When it does fully melt, we're left with a backyard resembling more of a muddy moonscape than a lawn, marred with the slog of a thousand paw prints, chicken prints, and kid prints, a few of my own, no matter how carefully I traipse, planted in the slick afternoons only to harden under the stars, thaw again when the sun comes to lantern. It's a repetitive process that takes the ground a month to reckon before the grass takes hold.

Meanwhile, I harp on the kids to take off their shoes when coming in, stack towels near the door to wipe the dogs' paws, which they hate, lament at how often they want out, lament the loud truck that made them want to burst out there, and how no one but me seems able to let them back in. It's no wonder the house never seems clean—always crimped by the remnants of the outdoors. All you can do is learn to accept your home will never be pristine. You can either spend loads of time re-cleaning it for a small moment of photo-ready tidiness or else let go of the need for a kind of perfection that will never last anyway. And while I subscribe to the belief we should leave behind every space better than we found it, this home is lived in. Soon enough there will come a time I'll wish the kids were here, unaware that they're one-part wrecking-ball, the dogs no longer sprite enough to chase the sound of every truck that booms past.

TO BLEND

When I was a woe-begone teen, I was obsessed with the poem "Invictus" by William Ernest Henley. The line "my head is bloody, but unbowed" gave me a kind of strength to weather whatever adversity needed weathering. Hereto, in whatever kind of shakedown I come across, I am steady as ground. Yet I didn't realize this so much until my divorce—the hardship of losing people in my life, known, trusted for years, and the struggles my ex and I had even communicating early on, so caught up in our emotions, and me unable to reckon the fact that my story, my truth, would never be told or heard—none of it bent me for too long.

Matt would remind me that it didn't matter what others believed, didn't matter if a story was wrong or any bit of truth would come. Let's walk barefoot on our lawn, he'd say, letting the ground *ground* us. Just let it all go, and so I did. No doubt his years of suffering from a disease that almost killed him gave him the authority to mind that outlook.

It was his ex that accidentally saved him, who'd stopped home for a moment to retrieve the guitar she'd forgotten and found him half-dead, unable to move his arm enough to reach the phone and dial 911, succumbed to sepsis. Having been rushed to the hospital and what followed—exploratory surgery that unveiled not only was he minutes from dying, but several feet of his colon were so deteriorated by Crohn's disease that much of it needed to be removed—refined his understanding of what mattered most. When I was frustrated with my ex in the early stages post-divorce when emotions ran high, or frustrated with kids, with my mind, or the garden not producing, or chickens dying, or the constant hiccup with projects—I grew from it yes, but I also learned from Matt to let it go. Nipped, clipped as a phantom limb, as the sucker that drinks all of a plant's nutrients, that takes up space but doesn't actually produce a damn thing. Being a bit more laid-back with how I reacted to things made it so the hard times to come didn't loom too large, didn't stick like a thorn in the backdrop of my thinking—that voice that likes to tell us what we can and cannot do.

The fall I turned 40 was a record book of heat. The air conditioning quit, the transmission in our car blew, and I wondered of the third as (Murphy's Law) these things always come in threes. Late September, the temperature was over 100 degrees and not a single company could take a look at our A/C for two weeks. For days I'd take the kids swimming at a nearby lake in the hopes that the autumn waters would stick, would keep them cool in the slow burn of evenings when the bricks of our home radiated with the high sun's heat, haunting our night's sleep.

We went to a posh French restaurant to celebrate my birthday, returned to a Hades-like home, wondering if we should just rent a hotel room, but dogs, chickens, the pride of sticking it out in unfavorable conditions—we stayed. I wondered if the nearby grocery store sold dry ice, imagined placing a chunk of it in front of our one fan, the cool fog of it enveloping our room like a stage. The upstairs was impossibly hot and no way any of us could sleep in the 100-degree broth of it, so I dug out an air mattress for the kids to sleep downstairs where it afforded at least 10 fewer degrees. Climbing the stairs to retrieve their pillows and sheets from their rooms, there's my husband, pantless, his black dress socks still hiked up his shins, vacuuming of all things, because he wanted to gift me at least a clean floor.

Of course, I couldn't sleep. At 3 a.m. I mixed a vodka tonic, extra ice, stripped down to my underwear, and headed outside to sit on the driveway. No cars would pass. No humans would stir and I would celebrate the milestone of my 40th alone with a Dogpatch moon, laughing at the absurdity of the heat, of my husband in his socks and boxers vacuuming in haste, old enough to know I couldn't ask for more.

And too, a gnawing at the periphery. Was I the 40-year-old I wanted to be? I'd gone to grad school in my early 20s to study creative writing, and though being a poet was not a career in itself as it would never afford enough to live on, I imagined I'd teach, would publish books, would always have a foot in the game. But the reality is I wasn't willing to move to follow a job to anywhere, USA to teach, and adjunct work unsustainable and spotty. I'd moved to Green Bay, WI for my ex's work and never found a community of writers. I got pregnant, headachy, and too tired to write or submit my work, and too strung thin for a year after Fisher was born to attempt then either. Toddler to raise, another pregnancy, then moving back to Iowa to be closer to family and more so, because I saw the potential in Des Moines. Then shortly after, divorce. We'd been together since high school and there's a wide berth in who you are and what you become. We'd changed, what we wanted changed, and I could only be thankful as it led to Matt.

I was riding the damn rollercoaster, too caught up with mom hood and moves and making something of this place to call myself a poet. But here I was on the precipice of a new decade, revealing the tick of time, and I thought; it's now or never. I stared into the swath of stars that night and vowed to bring back the part of me I'd set aside for so long. And that same tract of sky, some four months later, after I'd submitted my manuscript to a number of presses, poems sent to literary magazines, poems written, that sky would home the perfect X made of contrails at dawn. It would captivate me so that I'd take a photo to mark the omen it imbued, and later that day I would sign a contract for my book to be published. It would be released a few days shy of my 41st and I would stand then on my birthday at a bookstore down the road giving my first reading to mark the publication of my first book, awestruck at how it's never too late in becoming.

TO RENEW

Prior tenants had left junk piles throughout our yard—
an old doghouse filled with glass and roofing shingles,
tarp-covered piles of cracked, useless tires, bike frames,
bumpers, and a fax machine. Metal panels were strewn
about the lower level so that it looked like the site of a
plane crash. One 60s style swing set bent by the creek,
teeming with tetanus, and another duck-taped swing set
in the center of the backyard, the legs of it so rusted that
it would sway, threaten to topple over whenever the kids
would climb it. We took it down for safety, later to find
that old bucket of a playset was the kids' favorite thing.

We hauled it all out, burned it, or set it on the curb and
you'd be amazed how many metal scrappers would come
by to claim it, which they would haul to the local scrap
yard for any bit of cash it offered. Their truck axles about
ready to bust, the back end a jalopy of chain link fence,
patio furniture, pipes, and ductwork. When we gutted the
basement of its ceiling tracks, old insulation, and bits of
drywall and paneling, tossing all of it into the garage to be
dealt with later, one passing scrapper offered to take all of
it for us for 30 bucks. Heck of a deal for us. His wife didn't
say a word, but loaded up the interior of their van with
such precision there couldn't have been a peep of air left
in there, while he sat in the passenger seat and watched.

I hated the amount of waste our projects made, but we did recycle what we could, burned what would, or provide a source of income for those desperately needing it. We bought used when we could or found things. When storms would pass, the kids and I would walk the creek, fetching largish rocks I could use to border the landscaping in the front of the house. We had to have made over 100 trips one summer, up and down the steep hill, from the creek to the house, up and down the banks, sometimes hauling stones so large we could only carry one at a time. Over and over again, humping those rocks for me to edge in.

I found a sweet deal on flagstone at a landscape and nursery store nearby. They had it on closeout for a few hundred bucks, but the catch was that I'd need to move all 900 pounds of it myself. It took me a few hours, loading the back of the car with as much weight as it could carry safely, driving it home, carrying it piece by piece to the backyard where we'd planned a firepit, over and over until I got it all home. Another pattern emerges: tedium. When Matt got home, he suggested we move the stone and lay it out to discover how best it would fit and so I helped haul it again. Of course, we had to move them to dig down three inches, fill the space with paver base and tamp, which because we didn't own one, meant we set down a board and took turns stomping the ground flat. The stone, once again, put back.

Happenstance

I happened to pass a television tuned
to some nameless station and heard their
recounting: "I think we're doing
something more than what we thought"—
and you know the magic when
it happens, you say a word
and immediately after, that word
is sung in the song you forgot
you were listening to. Because
we do that a lot—sometimes
driving miles without remembering
if the stoplights we passed
were even green. And there it is,
some fishing line snared in our
hunger for known, and the fact
that the universe might be listening,
that we might be in on it too gives
the word repeated that extra punch

as if to say, see? When I heard
the line I didn't stop to watch
because I wanted it to mean
what I wanted—how even
in my mindless moving about—
vacuuming or folding laundry,
the bills to be paid or meetings to be met,
all the while as nostalgic petals
followed the time lapse of sun,
a lone cottonwood leaf dewed
to the sidewalk shaped a perfect heart,
that everything mattered, and
at any moment what I was doing
could be more than what I thought.

THE FLOCK

The ducks were a nightmare. The kids had seen one too many YouTube clips of adorable ducklings happily traipsing about and Matt liked the delicacy of their eggs, so we decided to bring home three from a farm posting I found on Craigslist. Their quack-quacking from the dog kennel in the back seat of the car as it kicked up gravel dust enroute was comical at first, but those quacks got really old, real fast.

Ducks need lots of water. I thought the kiddie pool, filled, would be enough to last for days. It wasn't. Those three had the water splashed out within hours. In hindsight, without a pond or the gumption to tend to a watering hole daily, I would never have bothered homing them. They are messy, loud, and their excrement watery and frequent. I'd divided up an area in the run using chicken wire to keep them separate from the chickens, but they managed to escape and within a few hours of having those ducks here we were already frantically chasing them through the yard, under bushes, across the neighbor's driveway, tracking them.

I woke the first morning with them here to eerie quiet—no chicken cluck, no ducks' incessant quacking—and peering out the window drew the strange wonder. The three dopey ducks stood along one side of the run and eight chickens lined the opposite, staring one another down as if daring the first to move. It was a standoff. I wasn't sure they would ever be able to coexist. It didn't matter. The chickens were let out to free range early that day so as to give them room to release the stress of those ducks, but I left the door open for their eventual afternoon visits to lay their eggs in the nesting box and to allow them their sanctuary should they need it, reinforcing the partition that meant to quarantine the three new ducks. I'm not sure how they managed to free themselves again, but clearly, we were not destined to have ducks in our flock.

I did search for them later, after we'd run errands and returned to find they had escaped again—brushing past poison ivy, the mark of the woods' cockleburs bedazzling my shirt, all those ankle-twisting roots and ruts, but I decided they weren't worth it. Plus, the welcome quiet of just the randomly established chicken cluck. We drained what was left in the kiddie pool and put it back in storage and we never looked back. I like to think they made it to a nearby pond or river, that they found mates and lived out their lives where they saw fit, wondering if the line of ducks splicing the sky overhead is in fact one of them or their kin, and wish them well.

We'd been lucky that so far all the chicks we'd acquired had turned out to be hens. Supposedly, there's an art to knowing, either by venting—which is to squeeze out the feces of a baby chick to open its anal vent enough to see if there's a small bump, deeming it male (even then not surefire as their parts look similar)—*or* inspecting the shape, the line of their wing feathers or the size of their comb. Some of these methods can only be done at certain stages of their growing, so when you're in the moment deciding which chicks to take home it's mostly a shot in the dark.

Casey Knott | 93

Turns out, three of the chickens we'd picked out from the farm in Jamaica, IA, where we'd tiptoed through a barn filled endlessly with the mosh pit of birds, were roosters. It was cute at first in an *I-can't-believe-they-actually-make-that-sound* kind of way. Astonishing, humorous, but does it ever get old. The thing is, roosters don't just "cock-a-doodle-do" at sunrise. Rather, they sound the alarm at all hours of the day. Often, to keep track of the hens because the one good deed they serve is to protect the flock. They crow when they see a hawk flying above or when the mail carrier delivers or because I'm probably on the other side of the kitchen window with a knife in hand chopping veggies, especially then, the jolt of their crowing hitching me like a surprise party.

The smallest of them, Candy, never bothered us at all and frankly, his tinny, toddler-esque crow only brought laughter. But we found him one morning, stiff dead on a perch, likely of a bad ticker. Lucy wasn't so bad—loud of course, but not aggressive. It was the Barred Rock named Rocky that raised all kinds of hell. There's that saying, breed like a rabbit—well this also applies to chickens. With so many predators about, roosters are forever trying to grow their flock. They're cocky, stand with their chests pumped high, wings drawn up so they resemble some bodybuilder flexing. And they mount those hens multiple times a day, using the long, sharp barbs on their back legs to clutch onto the hen's back feathers, their beaks to pull at the hen's neck, leaving her with bare spots on her body, the feathers unable to grow back in for the constant hook of the rooster. Rocky took to it like it was a full-time job.

A lot of friends ask if the eggs fertilized by roosters are edible or taste any different. A hen would have to go broody, whereby they sit on the eggs for 21 days to hatch a chick, otherwise if gathered daily there's no chance for a chick to grow so they are just as safe to eat and taste the same. We did have a couple of hens go broody one spring and decided to let it ride. As they sat keeping the eggs warm, only getting off of them once a day for a quick drink of water or a bite of food, the roosters became too much to handle.

Rocky had taken to chasing us. The moment any of us turned our back on him he would attack, sometimes running the length of the backyard to lunge at us, flashing his barbs and pecking us. It got to the point that even opening the main door to let the birds out to free range meant being followed by this roo and we had to walk backwards to the house to keep an eye on him. The kids were scared of him. The neighbor texted that her daughters had been chased across their yard by him and now refused to go outdoors. They had to go. I posted a Craigslist ad—free roosters—but rightfully, only Lucy was taken. I'd been honest about Rocky's aggression and not many would want to chance it.

The day I heard Phoenix screaming from the side of the house, that was it. I found her trembling and sobbing after having been chased by Rocky and all she said was that she wanted that chicken murdered. We had no other options. When Matt got home, he used a bent wire to hook the back leg of Mr. Rooster to suspend him upside down as chickens, like sharks, become docile then, and he shot its brain point blank with a pellet gun so that its death was a quick one. He did not get a burial but was tossed into the garbage can, where he put up a flutter, knocking from the inside for a good 10 minutes as his body kept flexing the final beat of his nerves. Dead and unaware of his final defiance. I never thought to eat him, nor any chicken we'd had that had been killed because we had named them.

TO BLEND

We don't just marry people, we marry their ailments, their people, their hopes and decisions, their defenses wrought from experience. Matt's ghosts were pretty clear. For the rest of his life he would never have a chance to sleep the night through again. Because Crohn's had shot his colon so much, he'd always live with a hole in his gut that connects to a plastic bag—sealed, itchy—to capture his body's waste, a bag that could only fit so much, that needed to be emptied at all hours or else it would run the risk of exploding. I try to empathize with Matt, tread light come morning, uncertain of his sleep, his mood, with the despondence of chronic pain, the embarrassment of our base nature on display. There are times he withdraws, having spent so much of his life alone with it and trying not to worry Elle, stinging with aches, and drained of energy when the disease comes knocking. And you never know when the disease will flare up, though stress has a great deal to do with it.

It was hard breaking in with Elle. She'd been fully in his care since she was two, through his surgeries and experimental medicines, all while he had to work full-time to support them. At one point he went back to night school to try to finish his degree, all while working 50-plus hours a week, caring for Elle, and with a body that constantly attacked itself, knocking him down. Elle would see her mom for a holiday, or a bit in the summer, but mostly she was there watching him stumble, recover, stumble, recover, again and again. She was more of a partner in those highs and lows than a daughter, and with her distrust of women, I decided I would need to be there for her on her own terms. I couldn't force our bond but hoped in time my being there would help ease her worry, some relief in knowing her dad was cared for and loved. In time, she saw that I was there—after school, during the rush of weekends, at her music concerts and school events, with Matt at the Mayo Clinic to establish his eligibility for a new drug, and the chance to have siblings, for better or worse, but siblings.

My own baggage, raw with newness, was not as easy to define. I had an edge about me to be sure, one that had carried me over in-the-scheme-of-things minor dustups, which I didn't see as a plot to replant, but more of a needed toughness to grow on. But in the end, I suppose it was a kind of armor that wound up being mended, remembering my mother's remark about how much softer I seemed after I connected with Matt. I was stubborn, too, and prone to anxiety, needy of assurance after the rejection that comes with divorce. And then the universe barged in, offering stubbornness and uncertainty. Yet it offered a different kind of assurance. I am loved, imperfection and all. And for as much as he has helped soften my definition, I know I have been Matt's purpose to carry on, no matter the pangs and setbacks. We are in it together.

Every family member knows I have a hill. Adorned with a scraggly looking crab apple tree, hearts and stars painted on its trunk, that overlooks our horseshoe drive, our woods, our winding road. I throw a quilt up there, write. I throw a quilt there and the five of us gather for a dessert of pie or the chance of meteor showers. Harvest moons, blue moons, the slow dance of July fireflies. The thing is this hill, the view it frames, reminds me of the life I wanted. Growing up, whenever I'd see a driveway, hidden for the trees, that seemed to undulate as a snake might, or an easy wind, I'd think—wouldn't that be someplace to live? And so it is that being there feels like a kind of wonder come true.

One evening out there, quilt top, looking skyward as if the words I needed to write were drawn in that space, I saw a hummingbird, perched and looking down at my blank page. At first, I thought how strange it was that I'd never seen a flightless hummingbird, and then realized this one watching me sat atop an electrical line that joined the street with our house, a line I'd never seen prior. For a moment I thought perhaps it was strung overnight, but when I asked Matt how long it'd been there, he looked like I was joking and responded, simply, *always?* For years I sat there believing the view, framing the reality I wanted—this unobstructed sky that I drew inspiration from, completely blind to whatever might challenge it. We are beautiful and strange that way—at one point, hopeful and at the same time, oblivious.

It reminded me of that sentiment, that privilege is when you think something isn't a problem because it's not a problem for you personally. It didn't mean I didn't deserve this view, having worked my knuckles to the bone, but rather that others, no matter how hard they trudged along might not ever have a chance of it. For as much as I'd been supportive of Matt's chronic disease, there was a side I would never understand. And yes, I believe we can always find common ground with others—we all experience the same emotions even if those feelings happen worlds apart, spurred on by a sea of things. It's impossible to walk in another's shoes when the world we view is our own design, so perhaps empathy isn't trying to walk there at all, but rather give them the space to speak and believe the words they use.

The best I could do for Matt was just be a source of love, not take his times of being rundown with his disease personal or fixable, to not say *Yeah, I've had my own aches and pains too.* It isn't about me even if it affects me. Only when we remove our own ego from the mix can we offer real support—his words, his space—I'd give him acceptance.

TO GROW

I grew up in a smallish town in Iowa where the tallest building hemming the skyline was the grain elevator. It was the stereotypical 80s upbringing—we kids were sent outdoors every day to play with the neighbor kids and weren't expected home until mealtime. And so we ran like a pack of wild dogs—hop-scotching over creeks, delighting in the fact that we could crawl under the street in one of its culverts, biking to the swimming pool most days where we'd chance the 50-foot diving board while Corey Hart, The Bangles, and Rick Springfield blared from the speakers and ride home hours later bleach-haired and sunburned; our parents just assuming we were fine and that we'd show. Latchkey kids. Hardly an adult around to watch us, keep us tame. We played red rover, freeze tag, and come twilight, chased fireflies, smearing their bioluminescence along our forearms so that we might also sing.

I grew up on Kraft and Oscar Meyer, on Campbells and Kellogg's. My parents never had a garden, never cooked with fresh vegetables or herbs, our meals concocted from cans via the supermarket shelf. A family favorite—tuna casserole, complete with canned tuna, canned soup, noodles, and crushed potato chips donning the top. It wasn't until I was out of the house and had to cook for myself, experimenting with recipe books, that I discovered raw garlic, fresh basil, parsley, and chives. I don't mean to knock my folks for the cuisine they imparted. I mention this only to say that I have not a lick of history in growing or cooking fresh food. No chickens running across the landscape of my childhood, and no idea how to farm, so my attempts here now are not derived from memory, but rather from intent.

My kids have no idea that I don't either. I enlisted their help with gardening from the start, and though the girls have grown less interested with age, they all relished in it. Even if they're not keen on eating the harvest of the seeds they might be planting, there was a sense that they were providing some good for the family, that they had a part in the whole and what they did mattered. Summer mornings, before the Iowa humidity would climb unbearably, we'd kneel in the dirt and pull weeds so that they wouldn't overrun our plantings and hijack all of the soil's nutrients, all the sun's energy.

It's lousy work, but necessary. Phoenix would ask how much longer we'd have to pluck; I'd answer, *As long as it takes*. She'd ask if we had to pull every single one and I'd say nope, we didn't. Too much perfection is unhealthy. The Japanese have a term for this mindset—Wabi-Sabi— which translates to "the art of imperfection." I came across the philosophy in my early 20s and adopted it readily. It's not an excuse to be lazy or haphazard but rather meant to embrace the fact that life is marred with rot, with decay, with cracks, chips, and broken hearts. To be marked is inevitable. Tree limbs, dormant, their leaves long dead, threading across a winter sky. Stretch marks from birthing kids. The worn. The sun-faded, the wind-tossed. To embrace that kind of wisdom is to accept the fact of life as a season—the birth and growth of it; the end.

So no, we don't pluck every last root of weed. A bird will drop any number of seeds as it wings across the expanse of our garden tomorrow, the next. Wind-fetched dandelions, creeping Charlie, cockleburs worn on the dog, planted anew, the constant expanding. There would always be more, so I limited our weeding to once every few weeks, some more, some less.

These kids do not have the same world that I did. I'll not be sending them out to make a day of it and assume the world will send them back. Things change. The risks are different. I'm in a city with city-esque issues and crime. But I can still give them a barefoot childhood—the dirt, the creek, the day spent outdoors, the fireflies come night. And especially the notion that there is a wildness to life that exists—one we shouldn't try to contain because it's the part that feeds our confidence the most. The discovery is in trial and error. Our upbringing should not insulate us from the fact that nothing is perfect. We need a bit of edginess to fully form ourselves. We need to know how to reckon with a world we might not always grasp. Perhaps one way is this—accept it and then work around it, weed and all.

Forgive Everyone Everything

The boney fox that killed
your favorite chicken
and dragged it to its den
where a litter of mouths nipped
and fed and fed.
Forgive the mosquito
for being a mosquito at
dusk after heavy rains.
Even the asshole
swerving his jacked-up truck
while he thumbs his phone,

a bumper sticker that reads
No Air Bags, We Die Like
Real Men slapped on his tail gate
just like his father used to do
to his mom.
Baby boomers, the 80s,
especially yourself back when.
Forgive us the day
so that we might be always arriving—
a worry no heavier
than a loose shirt.

THE FLOCK

The two broody chickens hatched our first homegrown baby chicks. We'd been checking daily after a few weeks of their sitting, anxious and uncertain of the exact 21-day mark. After the first week of their brooding, we'd snuck the eggs out from beneath the hens so we could candle them to make sure they were fertilized. If they weren't, they'd go rotten and would have to be disposed of. To candle them, it takes a dark room and a good flashlight to light up the egg. If you see a blob of white-like blood vessels and the outline of an embryo in the center of the egg, you'll know it's legit and can place them back under the hen.

Four new chickens and I had no idea what to do, aside from trying to hold them frequently so they'd be comfortable with human touch. They were born in nesting boxes some three feet high off the ground and were too little to move from there for the first couple of weeks. Featherless, they still spent much of their time under the hens to keep warm. The hens would take breaks now and then to get a drink or food, but they wouldn't bring anything for the chicks and I had no idea how much they'd need to eat. We put dishes of chicken crumbles in each of the nesting boxes, along with water, but every time I checked the hen was partaking instead, or the water had spilled. I just had to trust the hens would do the right thing, and they were, at the least, protective. They would puff their feathers and let loose a string of clucks in warning when we'd go to handle the chicks.

It's harrowing to see little four-inch-tall chicks in the coop with eight full size, hierarchy-testing birds, but the moms would use their wings to curtain the babies if other birds drew near. We were prepared to separate them from the main flock to keep them safe, as many people do, but we kept close watch and didn't see any of the other birds, despite their looming size difference, threaten the baby chicks. And the difference in size does not last long as the chicks grow quickly. By the sixth week, the chicks are considered to be teenagers, or pullets (if female) and cockerels (if male), at which time the mother hens abruptly abandon their maternal role. The pullets are left to fend for themselves and the hens have nothing more to do with them. If you didn't know, you wouldn't know which hens hatched which chicks, so final is their detachment. They become layers at around the 18-week mark, except for one of ours, who turned out to be a rooster.

Sure, we named them, but I can scarcely recall their names. You have so many chickens that come and go that it's hard to keep their names steady on the tongue. Of a few, I'm always sure—like Frenchie, the ones that hold on—but the rest become a kind of stand-in, cycling through and through as dogs and foxes and coyotes allow. And the kids took to providing such strange names like Canelle and Jaico and Soccer Ball; Fisher naming a multitude of them Lucy that we had to have been on Lucy III before I told him perhaps it wasn't the luckiest name—so that calling them by name became a kind of crapshoot. The variety of their breeds also escapes us with so many acquired from old barns, the farmers not giving one lick about their kind, and if they gave any a name, it would have been purely tactile, such as One Eye or Crooked Neck. Despite trying, no Barred Rocks could match the friendliness of our two originals—Jerry and Juice. We do our best, offering quality feed and space and protection, but still, they come and go. Phoenix believes our favorites are always the ones who get killed first because they trust the most and I don't doubt it.

TO RENEW

You hear something long enough and you can't help but believe it. In my earlier years I was told my hand wasn't steady enough for painting, wasn't allowed near a saw because my cuts wouldn't be accurate, and that I was, in short, too half-assed to get a job done right. But having installed kitchen cabinets, having painted the house inside and out, and all the number of projects that Matt and I had completed, I knew how wrong that earlier belief had been. I knew with enough research (YouTube watching), I could tackle just about anything. We remodeled our master bathroom, finally, top to bottom. Up till then, the shower in there had not only been nonworking, but it was a weird corner standing insert with a door that when swung open, narrowly missed the toilet.

I found tile at the Habitat for Humanity Re-Store, which saved us hundreds. Matt would cut the tiles and I would place and mortar. There's a kind of Zen to tile work. The scrape of the trowel raking mortar lines, the piping and smoothing of grout. I liked the room it gave my mind. I could follow the thread of a line for a poem, or I could think of nothing at all but the work in front of me, however tedious, it is transformative in the end. I liked having the tangible result of my hands' doing. We purchased a shower base and sliding door and we had ourselves a walk-in with room the length of the wall.

With our double-drywalled walls, it's impossible to find any pre-made cabinets to fit because they're made for standard building size, not our 1963 built home with that extra inch of drywall. A custom-built cabinet would have eaten too much of our budget, so I found a funky TV stand at World Market that fit the space well. We cut a hole in the top, installed a vessel sink and had a one-of-a-kind vanity. It's astounding the cost for most contracted bathroom remodels—an average of 10 grand, and a good chunk of that cost is labor. But when you do it yourself, when you shop around for the best prices, when you find alternative products and make them your own, the cost is much less. Our bathroom remodel (complete with a new shower pan and faucet, sliding doors, tile surround, tile floor, lighting, fan, vanity and sink, patched in drywall, and commode) was under two grand. All of this to say whatever your budget, it's possible to make it work.

The law of thermodynamics states that everything in nature has energy that can't be created or eliminated, only exchanged. To put it simply, all life is an exchange of energy. It's why laughter is contagious, it's why a wanted hug feels good or someone's foul mood can adversely affect yours. I've noticed too how that energy, like a ghost, can linger. We don't just leave behind secreted nudie pictures and empty beer cans stashed in attics, long forgotten. We don't just clear the tools from a finished job and assume the space is ready, new.

In college, I studied abroad in the Netherlands for a semester—took the train to Germany, France, Italy, the Czech Republic, and a dozen more—places with more recorded history than ours. At the Dachau concentration camp near Munich, the air feels heavier among all the photographs of skeletal-thin prisoners, imprinting the atrocities so that we never forget. The Gothic buildings in Prague, spared in the bombing raids of WWII give testimony to what all of Europe might have been. Every cobblestone street, despite the cars, echoes still the former click of horses. The air so pocked with history, seems charged. We can't help but leave our mark.

When it came to the downstairs bathroom remodel, a simple facelift of paint, new floor and small vanity purchased new, but one which still required chiseling away the old tile at the base of the walls, some sad news planted there. My ex rang to tell me about an old beloved neighbor from our stint in Green Bay, WI—the neighbor who used to have one of his five kids stopover on a random Saturday to deliver oven-sprung brownies or a batch of homemade salsa, a bag of chips to try it with even because he thought of everything like that; always two steps ahead—died in a car accident. His second youngest kid, at 16 and driving him, also gone, on Mother's Day no less. She hadn't paused at the stop sign at the intersection of some rural road long enough, hadn't seen the truck barreling through. That mom, those siblings, and the waves of memories with them struck me right there on the bathroom floor so that I can't go in there now without remembering them—if even a whisper of a thought, as if whatever happens, happens not just to us but the ground we sit on, the air we exhale into. This life forever making itself known.

TO BLEND

I was giddy in love with Matt for some time. The kind where his presence aligned with mine seemed charged, atmospheric, as if our bodies were magnetized, were meant to connect. That honeymoon phase. In the beginning we couldn't stand to be apart. And then, like a film scene where the camera zooms out to reveal the larger picture, we came to know all the ways in which we differed. For one, if you were to look up the definition of patience, you'd find its root meaning was everything my husband stood for. His signature phrase might be, "It'll come when the time is right." He's late all of the time. I probably spend a solid week every year just waiting on him, the kids and I restless in the car while he tends to a number of last-minute things.

His attitude is that things root and bloom when they are meant to. That's all well and good, but my root sense is that things happen because you work to make them happen. Not that he hasn't busted it and worked hard and not that I haven't seen the beauty in the clouds spilling their guts across my bit of lawn, the best part of my day spent atop a quilt watching the planets turn—just that our approach to things is vastly different. Yet I see it as a necessary pairing. He might be the kindling and I the fire—either way, the prairie burns so as to gather its nutrients and renew.

So he has a bit of a laissez-faire attitude. Our dreams are the same, but our methods are nothing alike. But it turns out that his way of slowing and believing in things is the slow burn I needed. There's a line in Neil Young's song, "Love is a Rose," that Matt would recite to me—how love "only grows when it's on the vine"—when I was heady with uncertainty, post-divorce, so as to assure me he'd be sticking around. Those lyrics embody pretty well everything now.

We do what we can with what time and space and ability allows, but everything is transitory—nothing owned. It's like trying to capture a seascape sunset with a camera—the moment lost to finding the ideal placement, to setting the equipment, to standing behind the lens. To try too hard to make it just so is to miss the stirrings out of frame—the violet hues aflame behind, the awed quiet of the child beside. To try too hard is to render it unauthentic. As Thoreau once remarked, "All good things are wild and free," which I've come to know is true in gardening, in homing, in loving. Sometimes the best we can do is let things be, to see it grow, and to just sit in that enjoyment.

Having kids that divide their time between two homes, accepting that sense of vulnerability is crucial. Being a mom is hard enough with all the guilt we place on ourselves, wondering if we're doing enough and doing a good job raising these tiny humans to be healthy, adjusted adults. Add the fact that your kids aren't home to tuck in every night means two things happen: guilt for the wonder of what you're missing and also the unintended but real shaming from other, non-divorced parents when they imply that they don't have the luxury of catching a show on a Tuesday night because they always have their kids. It's weird. And it can be a draw to go over the top planning activities for the kids to make up for lost time, especially when they start rattling off all the things they did at their other house or that they always call their dad when they're with you but never you when they're not. Those are breeding grounds for the sense that parenting is a competition, intended or not.

Here's the thing—it gets easier in time. You survive when you stop taking things personally, you thrive when you remind yourself of the simple truth that no one and nothing will ever change the fact that *you're* their mom. When they're not with me I utilize that space to delve, uninhibited, in my writing because I am not one thing in this world. And I also have this midwestern stubbornness that wouldn't otherwise give myself a break for the world is full of people with harder roads to tow. How can allowing myself time to be quiet and soulful, free of guilt and need, not ultimately make me a better example for the kids—one that encourages them to nurture their dreams?

It's not a competition. What the kids need is stable ground, love, and support, and for me to be me—the very best version. Hadn't all of our plantings given some roots? Hadn't our weathering with tired muscles and busted knuckles and lessons not made us more sure and ready? Hadn't we found each other, dreamed up this home? Hadn't I discovered that life is one part what you strive for and one part what you accept? I'm not going to lie. It feels like a gut punch or a spirit deflating when the kids come home with stories that include their "other mom," but it changes nothing. I am their mom and the fact that they have more people in the world to look after them and love them only serves to enrich their sense of belonging here in the world. Sometimes I have to remind myself of this, but mostly I trust that it is so like a wind billowing open, pushing the sails through waters chartered or no.

TO GROW

I never could get cherry trees to grow. I tried, for three seasons, to plant them on my favorite hillside, widening the dug holes each time, filling them with a blend of worm casings and organic soil, plenty of peat moss and gypsum to help break down the clay-prone soil, give the roots a fighting chance to tunnel underground. I fenced them off so that deer couldn't reach the young shoots, which they liked to devour, and with enough berth so the bucks couldn't use the trunks as a scratching post in the early fall to rub away the velvet layer of their antlers, or come rut, to mark their territory. I watered weekly when the rain didn't provide and covered their crowns with netting when the Japanese beetles advanced, hellbent on their leaves. Still, the beetles found their way in or else the winter freeze would prove too much. At a certain point you've got to cut things loose. Not everything will grow well here and when I discover what does do well, that's what I nurture. Like the ducks, cherry trees were not for us.

We have an abundance of berries—raspberry, blue, and black of varying varieties. We've planted these at multiple locations throughout the property with the idea that one might be able to enjoy an edible walk. The berries seem to flourish wherever they are—hillside, partial shade, or open to the full lick of sun. The land seems especially fond of blackberries as they grow in such numbers that the only way to avoid having the fresh hundreds uneaten go to waste is to cook them down and make a jam, my favorite concocted with the herbaceous balance of fresh sage and the bright citrus of orange peel, a dozen jars-worth that will line our basement shelves.

The only edible mushrooms that sprout willy-nilly in our woods are turkey tail, chicken of the woods, and the occasional, prized morel, none of which we had a hand in. The ones we did attempt—shitake, oyster, chanterelles—which are supposedly easy to grow, did not, in fact yield anything. Given our variety of trees and the fair amount of decaying limbs that dot our woods, given the north and west sloping land, the creek, the sunny flat, the compost pile brimming with the chickens' fertilizer and coop-tossed straw, we imagined it an ideal habitat for any number of fungi. Even following the directions: drilling holes in freshly cut hardwood logs, hammering in the mushroom plug spawns purchased from a reputable online mushroom store, sealing them, and keeping the inoculated logs moist, elevated, and shaded, we got nothing, as if the land here already decided itself.

When the turkey tail and chicken of the woods sprout, they do so in numbers so large that we dehydrate slices of them to store in jars for future dinners. The morels we find are so rare and spotty—maybe three or four a season—that we pick and cook them immediately. We tried for more—a bag of their spores kept in sawdust that we stored and soaked and sprinkled on the sun-filtered hill, but years later not a one has popped there. Now, instead of attempting to grow any, we walk the property, forage the gift of its offerings. Some things better left for the land's bidding—not like we ever had a say.

I'm sure the neighbors think we're strange, how often one of them might pass and we're either sitting atop the driveway or on my hill or we're picking berries from our random patches, searching for mushrooms, strolling with the dogs, or else exploring the creek. I've always believed one of the best things I could do for my kids was to teach them the simple fancy of just being in nature, of looking into the sky or sitting with the scent and sound of woods and marveling, or *not* marveling. I wanted to instill in them the contentment of these seemingly simple spaces so that it would be a kind of home, a refuge in the eventual ups and downs they'll face. For them to find the beauty in a small thing—the greening of spring, the light through the trees, sky gazing, a forest hike—will hopefully root them to what matters most. It's not the material things; it's the very nature that surrounds us, that's in us, that will give a sense of meaning. I don't want them to have to look far to find something that instills awe. It's not about giving them all the latest technology, gadgets, fads, things, and trips, all of which will only make them more expectant. I'm teaching them to find satisfaction in a moment, not for purchase. Less is more.

Some nights out with flashlights, mud boots on, we'll find ourselves caught in a passing car's headlights where the road curves and we're alit in the creek. Do they wonder of us? Do they notice at all? Is it strange that a 40-some-year-old woman is ankle-deep in the creek by the road on a Saturday night? I'll be there at midnight, looking at the minnows in the few pools deep enough to hold them, a melancholy for their small chance to grow. What sustains them? I never want to be an old lady who doesn't walk in a creek.

Old and in the Way

Your father calls—
it's your grandfather's hip this time,
busted like a 5-point star.

He tells you there's no dignity
in being that old
and if he gets there in diapers

looking into space unaware
of the season or if
the 4th of July was the 6th of June

to take him out to the woods
and tie him to a tree for wild
animals to finish.

You'll tell him that would be a slow
and painful way to go—
better to find a cliff

and push the weight of him.
Of course he'll agree.
His father is angry with the sunrise

and even the setting. Besides,
your dog is old and in the way
and you feel bad about it some.

This is supposed to be
the twilight, but the dog
keeps shitting in the house

and placing his old bones
in front of every door you propose.
His eyes crust

and he smells of dead things.
Still, every now
and then he wags his tail

and you can see far
into the foggy marble
of his eyes that he sees you

and he'd come
running if he could—
some flicker of what

he once was—and it's a thin line
he casts to you,
and it's a selfish thing,
being alive.

THE FLOCK

To farm, however small, is to know death. The new rooster was gentle (in the sense that he never would attack us) so we kept him on. The hens would probably disagree on the gentle assessment given their constant mating. When a few of the hens went broody, we decided to let them keep a small clutch of eggs to brood over. Five more chicks would hatch. The rooster turned out to be surprisingly sweet. I'd see him give the hens a break—a chance to stretch their wings and scratch in the woods while he'd stand guard over the littles.

We had 15 birds in total at that point, hard to keep track of names and whereabouts. Headcounts with a flashlight shone on the perches come sundown, sometimes one or two short. We never knew exactly what snatched them, here and there a smattering of feathers, always small, and always it happened when we brought the dog inside to eat as if we were the recipients of a stakeout. Of the three chicks that survived that summer, one was another roo. At first, it was easy, quaint even. The dad was in charge and the younger rooster didn't challenge. But as it grew, so did its need to become a legit rooster, to have its share of the hens and test the hierarchy.

Ironically, one evening, when Matt had just finished remarking how lucky we were that the roosters got along so well, a cock fight began. The dad was much larger, but that also meant he was slower and could not keep up with the ware of the chase. They ran figure eights around the house and the backwoods, squaring off now and then, using their barbs and beaks to pounce on the other and draw blood, mostly the younger one achieving the most prominent points of injury. There'd be no Rocky Balboa comeback—this dad was not going to win.

Anyone who's ever tried chasing a chicken knows the only result is you end up looking like you're chasing wind. They're almost impossible to catch. When it comes to a cockfight, there isn't a thing you can do but let them wear themselves down. After that, the best option is to kill them, as once you have a roo that fights to the death, he'll do the same to any hen that stepped out of line. And the larger rooster, the dad, bleeding from his crown was too injured to survive, so Matt hooked them by their legs, one by one, set their necks atop our wood chopping block and quickly, swiftly, axed their heads.

Some might have relished a dinner of coq au vin, which prefers the tough meat of a rooster cooked over low heat and slowly to render it tender. Though I was never keen on eating our hens, I did consider making a meal with those troublesome roosters, named or not, but all of this happened on a weeknight after a long day of work and we were not prepared to process a chicken to eat (plucking feathers, removing innards) so those roosters' final resting was again, the garbage can.

It was heartbreaking to watch Neko's age start to prevent him from what he loved most—playing fetch or chasing squirrels. He had arthritis in his hips for some years but no pill, no supplement relieved it at this advanced stage. His eyes were milky, his body lumpy. Some days he couldn't manage the stairs and Matt would have to carry him up and down, indoors and out. The vet believed he had dementia—sometimes I'd come home to him standing in the corner of a room trying to go forward, head-butting the wall as if he'd forgotten how to turn around.

A hard decision, to end your dog's life, but he was 14 and in such poor condition, it seemed the kindest way. For months I'd come home half expecting to find him dead, sometimes even hoping to so that I wouldn't have to be the one to decide his fate. But the best I could do for him was give him the dignity of a peaceful end with me at his side and stop his suffering. I told the kids, and neighbors—everyone had a chance to say goodbye, and we had one last day to celebrate old Neko. Wouldn't you know, that dog leaped from the car when we arrived at the lake to let him swim, like his joints weren't a knotted mass of ache, like he didn't just shit in the house every night for the last few months because he couldn't control his bowels or else remember where to relieve himself. It was strange, watching him swim so carefree and knowing in a few hours I'd be holding his paw while he crossed. I like to think he was gifting me the memory of the dog he was in his prime—a kind of nod to the good life he'd had.

Fisher and I dug his grave in the woods while Neko sat sunning on the hill. We choked on our sobs, smacking away mosquitos and flies out with the sky threatening rain. And later, the family gone, I drove as slow as I could down the road to the vet, questioning if what I was doing was the right thing, trying to eke out precious minutes with my dog. Neko, who'd seen me mother, who saw me through the divorce and rebirth—my steady companion through it all, I held his paw and cooed to him, good boy that he was, all the while his light dimmed.

There are coincidences but I believe there are also signs. The X mark in the sky, the feathers, and also the day after burying Neko, when I sat on the picnic table out back, teary and raw in my grief, a robin landed and swiftly hopped the short distance between us until it was a single stride away. Neither of us stirred and neither looked away for a long beat of time. Neko loved those birds. He'd watch them in the yard and lie stone still so that they'd draw close to him. Something told me there was an afterlife for dogs.

TO RENEW

So many storms have come through—the weight of ice in winter, the high humidity of July thunderstorms. We're on our third chainsaw now, as almost half of our acreage is wooded and the work to clear its' debris is constant. We've had trees fall akimbo all across our property. Some we leave to rot, to house wildlife and fungi, but many we've had to process—twigs, branches picked off and larger limbs and trunks cut in two-foot lengths to then be split with an axe, or on occasion, a rented log splitter. To hear a tree fall in the woods sounds like a spooked deer stirring a bush, or the sudden breeze kicked up before a storm hits, the thwack of a wooden bat as it strikes a ball—to hear a tree fall is surprisingly that small. I'm forever hauling lumber, stacking lumber, fetching it for fires.

Before the last of it is cleared you can bet another tree will fall. And too, unless you spend the extra time and money to have the stump left behind ground down to the roots, terminating it fully, that tree, like an unwanted thought, will haunt, will nag at you. For years the shoots not just around its base but the span of its canopy's projection—branch spoked or wind ridden—will grow anew hundreds more trees if you let it. Nothing ever gets gone. We hack at the shoots with log choppers and pruners, depending on their size, to halt their cropping, trying to claim this space as ours, a bespoke irony. I'm sure we don't get them all. In time, the space we inhabit the most in our crossings will remain bare enough and the offshoots will season themselves enough as a tree that we'll let it be to thrive, only to threaten again someday.

We had our first Derecho storm over a year back and in all my life I'd never known such a weather event existed, but I imagine with climate change, such inland hurricanes will become more prolific. The day it hit was the day we were moving Elle into a new college apartment up in Ames. She was all packed up and Matt was just about to head out to pick up the U-Haul when the sky charcoaled and the rains came in sideways, dropping hail and dumping torrential rains. In no time our largest tree out front—a 150-foot buckeye—was snapped at the base and sent toppling into the electric line, effectively cutting off our power and barricading the driveway. The rest of the yard was a Jackson Pollock of branches and leaves.

We were quite lucky, we discovered, as we drove the packed U-Haul to Ames just a few hours after the storm, meeting roadblock after roadblock—downed trees, barn roofs, and overturned semis making some areas impassable. The whole city of Ames was without power and we hadn't thought to pack food or extra flashlights, not realizing the extent of the storm up there as well. Imagine unpacking your college kid in their new dark and muggy apartment with your youngest shining a lone flashlight down a hallway to help you see and then, after attempting to find a restaurant open and failing, leaving them without a meal or electricity to begin their new journey.

You have those rubberneckers after storms—the ones that drive around not looking to help, but to see with their own eyes the damage. Others drive by looking to profit.

So many of them passed along our road remarking the fate of our buckeye tree, shaking their heads at the sight as if to say, *Sucks to be you*. I heard a young girl say to her mom as they passed, windows down, *I feel bad for that tree*. I did too. It wasn't that our power was shot and wouldn't be fixed for unknown days, it was the space it left behind that ached. There's a kind of reversal in nature—to see a young tree matchsticked on the lawn is nowhere near as felt as a large, 100-year-old tree on its side. To think of all the birds it served, the air it made, the history it shaded and cooled was to invoke a kind of reverence.

I felt this pull to honor that tree, so I lit a sage bundle and smudged all along its trunk, the point of its splitting. Because I do that a lot—make a ritual so as to imbue a deeper sense of meaning, to make a day more heartfelt than the routines of living would normally allow. I thanked the tree for all it offered and in my own way I blessed the space it left behind. The next day a city worker stopped and told us not to worry about the tree as they'd be by the next to clear it for us, and the energy company had a boost of help from workers several states aside who came to help with the massive power outages. We were told ours would be up within two days. In the meantime, our neighbors, helpful and kind as always, offered freezer space so we wouldn't have to toss all of our food and cell phone charging and internet passwords so we could stay connected. Magic happens.

TO BLEND

I knew these roads as a kid. Though I grew up in a town 20 miles south of here, it was an area I'd happened by now and again from the backseat of my parent's sedan. Back then, I imagined I'd live along the ocean in California, adopt a dog, and learn to surf. All I wanted at that age was to be far away from what I knew. How could we ever imagine what we'd become?

One of the best things anyone has said to me was when I went in for a sports physical in high school. The doctor had asked me what I wanted to be when I grew up and I said I had no idea. When he replied with, *I don't either*, I was stunned. If this man could go through years of med school, a residency, at least a decade in of work there, and still wonder if he'd chosen correctly, then maybe it was okay not to know what I wanted to do and maybe I never would. Maybe it was that my purpose would evolve as I did. I can't think of a better message to tell a young teen—don't get too caught up, too worried in planning a future you can't know. Try things. Experiment. See the world. Believe that the right things will take shape in time.

A career was not the definition of your character, a career was not the most important thing. For a young poet's heart, could those have been larger wings? As natural as the conversation starter *how about this weather?* or *what do you do?* is one I'm also guilty of, though I despise it. Mostly because when I reply, *I am a poet*, most don't take it seriously as it's not a high-earning endeavor and, in many minds, not an actual profession. Even though most of us are wishing away the time between weekends and holidays when we can do what we really choose. So, it's become that what I do to make an income is second rate to me as whatever my career, it is only to buy me time to do the work that is meaningful—being with my kids and writing poetry—and both perhaps the most thankless jobs of all. No matter.

I could never have guessed back then from the backseat of my parent's car that I would ever live here. It's as if the future is always riding shotgun with our present—*see that, you in 10 years*. I should have known that was why my old place, pre-divorce, felt so transitory and strange. I remember driving home from this or that in those days—a grocery store run, the kids to preschool—and thinking, *Is this my life?* (Cue The Talking Heads). Driving past the house on the corner with its tree stump carved into an eagle, talons pinched into a fish, or that house that moves their trashcans to the curb always a day late for collection, same dip in the road, same view. I couldn't imagine that route being forever, and then it wasn't.

And now, all I can think is give me one more day, at least, on this one. Sitting here on my favorite hill with Matt and the stars, the kids tucked in and pretending sleep. The only talk of weather is when I ask him where this wind comes from, to which he pauses, says he doesn't know how to answer that. I love him all over again for the care he takes with his words. Dear future, let him be here, the both of us old and new to each other. Let our kids grow into their lives. That's all I ask for.

Exposed

It's a small thing—
a nail here, a shim
boned in, a few coats of paint,
mortar troweled flat as
today's mail—none of it
worth opening—before
the tile is laid.
Before the stain you must
sand the wood drum thin.
You must sweep at once,
measure twice,
prepare for the worst—
cock-eyed screws with heads
shredded, a busted pipe,
the warped boards,
the black mold, the stitches
and the smashed digits,
and, inevitably, always one
problem begets another as
one kink in a slinky and all of it
stops—

that's how small.
You cuss a lot,
you thank the gods for Duck Tape,
cross your fingers, crack
a cold beer. And then
you realize the hardest part
is living with what you've done—
a line so thin
you could hardly imagine moving
freely again,
though you must—

the bruises yellow, fade,
the pin tuck of splinters become
part of your skin
and no one but you notices
the faults hidden
in your work—
all those mismatched lines,
the incidental nicks
in the wall you left. Somehow
that's all you see.

TO RENEW

We remodeled our main bathroom damn near to the studs. With nary another project to finance, we splurged on a custom fit vanity and countertop. Still, I found deals on the tile (matte white subway tile for the shower surround, and hexagon tiles for the floor), tub, faucets, fixtures, and a mirror by comparing prices across several sites. The first rule of remodeling is that there will always be a hitch, some problem uncovered in the demolition that throws off the timeline and gouges the cost. You can plan all you want, and you should, but there will always be an unforeseen hiccup.

That old neighbor back in Green Bay, now gone, loved to share the story about how his bathroom faucet began to leak and upon hiring out for repairs ended up needing to tear out the entire bathroom and remodeling it ground up because *nothing is as it seems*. One of those, *If you're changing out this, you might as well do that as well*, and then the list repeats and before you know it, everything is changed. Beyond his nuggets of truth were his random acts of kindness. Sometimes he'd drop off a DVD or two just because, just in case we hadn't seen them. Perhaps we'd enjoy anything from *The Little Rascals* to the complete music video collection of Michael Jackson, the latter of which I showed to one-year-old Fisher (sans *Thriller*) and who became so enchanted that I'd hand it over like a Bible to anyone who babysat him with the instruction that if he got fussy to put the disc on as a kind of balm. It always worked. And here that man was, an echo still.

A sledgehammer and chisel on the original 1960s beige shower tile revealed mold in the wall insulation and on a few studs. Easy enough to tear out and replace. We managed to tilt the tub itself (a cast iron low rider in a peachy, putrid shade) to its side but knew we were not equipped with the muscle to remove it from the bathroom and haul it outside. I called a junk hauling company who came to assess and quote the cost of its removal, as well as the demoed debris from the old vanity and busted floor and wall tile.

Yet, when the workers hired for the heavy lifting came, they took one look and said there was no way they were getting that tub out of the bathroom, let alone down the hall and staircase, without causing damage to the house (and their backs) along the way. Typical for the higher ups in a business to be completely out of touch with the reality the workers they manage face, their minds set merely on profit and not the legitimacy of the work it might take. The guys were kind about it, and I sincerely felt for their predicament as it was the same way I felt and why I called for help in the first place. They offered to haul away the debris we had managed to move out (piled surreptitiously in the catch all of the garage) free of cost and gave me some numbers of other places to try and then, trailer loaded with our unrecyclable, unusable debris, managed to get their truck stuck in the snow pile shoveled aside the driveway which meant a full on hour of shoveling, salting, wedging plywood strips under the tires to get those poor guys free.

No one else would take the job, but one guy offered some sage advice—the only way to get a cast iron tub out properly was to do it bit by bit. I watched a handful of YouTube videos for "cast iron tub demolition" and discovered that once you find the weak point (which means taking a sledgehammer to its side and delivering it till you get the first crack) the rest of it can be broken up quite easy. I busted my finger on my second go then left it for Matt and Fisher to blow at that evening and bit by bit that tub was disembodied and taken to the garage. For weeks we would feed the shards of its remains, bit by bit, to the trash receptacle.

And there I am zenning out with tile work. Spreading mortar, placing each piece just so like some grown up version of Tetris, the satisfaction of seeing its completion row by row. But the strangest thing happens when the work is complete and you have a hand in making it so—knowing where the subfloor was rotted and replaced, knowing the points, where tired and worn thin, the nail you were driving might not have hit its mark, and that troweled mortar beneath that tile is only so thick—you can't help but tread lightly.

Seeing the work that goes into making a thing gives it a kind of resonance that hits close. I tiptoe through the landscape I've created because I know how delicate it is. Not because I don't trust the work, but rather I see it so closely and want to make it last. When you hang the cupboards that will eventually house several pounds of ceramic plates and dishware and realize the only thing holding it there is a strip of metal fastened with a row of screws teething at the wall, I guarantee every time you go to place the clean dishes on its shelves you will marvel at such suspended weight.

My grandmother died suddenly one late summer years ago, having just returned to Arizona after visiting and camping in Northeast Iowa where a tick, no larger than a crumb, infected her with Rocky Mountain Spotted Fever. The things that mark us are small yet enveloping. Lucky for anything that survives. After she passed, my grandfather, A.G. (which isn't an abbreviation of anything), subsisted on McDonald's hamburgers that he discovered were four for $1 on Thursday; so he'd stock his freezer and have a ready-made meal for the week. He got by; met and married a fellow widow and the two of them were companions for some years before she succumbed to dementia. Alone again but content to synching his days to the routine of walking, lunching, catching a show on the boob tube.

When he turned 90, I called him at the nursing home where he'd been relegated to after falling and breaking his hip. His memory had been dwindling and I admit I hadn't been the most dutiful granddaughter, which was in part because he'd been living two states away and I wasn't sure he remembered me—at least it seemed so in the scattered phone calls we shared as he inched his way through the tail end of his 80s. He'd often repeat the question, *Who is this?* and I'd have to remind him that I was his son's daughter.

As a child I remember running my palm over the soft edge of his buzz cut, which reminded me of a teddy bear, and paradoxically, his penchant for cactus plants that prodded from every nook of my grandparents' trailer. He'd once owned the DX gas station in their small southern Iowa town, which my father was obligated to work in, forgoing any extracurricular activities.

A.G. had been a serious alcoholic during my father's formative years and he wasn't ashamed to mention it because his sobriety was a measure of pride. He'd tell me how one day he woke up and decided to never drink again and so he didn't, end of story—a kind of scrappy attitude of one who was born and grew up through The Great Depression, dropping out of school after 8th grade to work on a much-needed paycheck. Through retirement, Grandpa fished, told jokes, went for walks, lived simply.

But when I reached out on his 90th birthday with my well wishes and congrats, I was hoping for wisdom— *What's the secret to pushing nine decades?* His answer was this: *You know, Casey, if someone orders a pizza, I don't care what the toppings are, I'll eat.* I don't know what I was expecting, but it wasn't this. Surely a life worn in would reveal some semi-profound insight. For a beat I had to swallow my disappointment as we continued our chat— him asking again to whom he was talking, him telling me he seemed mostly to think about his parents, his early years, as if the ending truly was in the beginning.

It wasn't until the phone call ended that I reflected on his lackluster words and realized what he implied in his round-about, humorous way—*Granddaughter, don't get hung up, dying on innumerable hills as if there's just one way. Rather, go with the flow. Be light. Be soft and adaptable. Sustain yourself with whatever bone (or pizza topping) the world tosses at you.* His metaphor meant this—don't stress too much. It's the worry that digs the grave.

THE FLOCK

Just before the first COVID-19 lockdown, when we knew what we didn't, Matt, the kids, and I ventured out to the local farm and country store to pick out baby chicks, and apparently a turkey, thinking what better time to hand raise chicks and also to provide the kids a source of entertainment, hoping to steer their minds from the wonder of their schools being shuttered. Impossible to know what we were facing. We had 24 hours to move Elle from her college dorm and bring her home after having spent just one semester away. We knew nothing of turkeys either but on a whim decided to see what it might mean, perhaps to make the sting of the unknown less of a worry and more of an opportunity to behold something new.

We got many of our chicks at the farm store in the spring when they're full of hatchlings for sale, only seeking out Craigslist in the off-season when an attack on our flock deemed it necessary. The eight chickens we had currently were more skittish, aside from our old holdout, Frenchie, who'd still let us get close enough to pet or pick up. We longed for the sweet docile nature of our original flock, so for the first six weeks of lockdown we lived with four baby chicks—two Sapphire Olive Eggers, a Buckeye, and an Amberlink, and one broad breasted turkey that we named Thumper.

We sat with those birds every day, and when they weren't snuggled in the crook of our neck or purring in the cup of our palms, they were cozy in a dog kennel set in our living room. At six weeks, the smell too overpowering, we moved them to the garage as they still weren't ready to be with the full-grown flock and we wanted to protect them for as long as possible. Thumper was double the size of the young chickens and with her long neck and legs resembled a miniature ostrich.

I wondered how they'd all get along, if she felt herself an outsider, remembering that quiet morning standoff with the ducks. I'd noticed too how our chickens often paired off with similar looking birds in our flock—the brown feathered chickens teaming up with other browns, the light ones in a troupe their own. Though they're a flock, they don't necessarily stick around in one big group but rather wander here and there foraging or sunning in pairs, and occasionally some complete loners who did their own thing entirely—not unlike middle schoolers at their first dance. But they were all drawn to Thumper. Even when it came time to move them fully to the coop, after their two-week stint in the garage, and introducing them in brief sessions with the older birds, not one, not even Frenchie tested her. Thumper didn't make a move to challenge the totem pole—it just came to be natural that she was respected, like the cool older cousin you wanted to impress when you were a kid.

She grew twice the size of the chickens. She'd fly up and perch on the top of the coop or trampoline enclosure as if to avoid being locked up at night. In the wild, turkeys perch in trees, but we wanted her in and safe from racoons at night, so it took luring her to come down with her favorite treat—watermelon. Still, come day I worried for her roaming. Sure, she'd greet us at the door with her loud yelps, she'd sit at my feet, steal sips of wine from my glass, or peck at the pen in my hand as I wrote and watch me side-eyed like she was trying to read my mind. When the mail carrier would come, she'd follow me to the road to retrieve it like some guard dog. In fact, she hung with the dog a lot. If Luna was napping in the shade, Thumper would lie next to her and purr. If Luna was running the property line, barking at passing cars or pedestrians in warning, Thumper would flap her wings, squawk, and run like a banshee as backup on that line. Turkeys can run up to 12 miles per hour and have quite a range of vocalizations, from yelping when excited and purring when content, to a kind of "kee-kee" sound they make to get your attention. And they don't just express themselves in sound, they also change the color of their heads (which are nearly featherless) to show their mood, from red to blue to white—the more intense the color, the more intense their mood. Thumper's coloring usually comes in white or pink and the few times that blue has splashed across her she was charging towards the road alongside the dog, a kind of warning.

I've had neighbors stop to tell me how Thumper wards them off, puffing her feathers and staring them down whenever they happen to be by on a walk. She's made a name for herself here, but the problem is she sometimes hits the road. I've driven home to find her stopping traffic. At first curious as to why some car was stopped in the street, blocking the access to my driveway, only to realize it'd be Thumper out there—squared off with the front bumper of a vehicle, daring them to pass our place.

The remarkable thing is she knows her name. If I venture outdoors and don't immediately see her all I have to do is holler, *Thumper!* and she'll answer with her usual loud chirps until I find her or until she finds me. If she's across the yard all it takes is to hear her name and she'll come running. We are all fascinated and smitten with her. Of the four chickens we hand raised that spring, the two Olive Eggers—dubbed Mento and Olive—were the favorite. Like Thumper, they were keen to greet us at the door whenever we stepped out, following us, eating from our hands, and curling up on the quilt we spread out frequently for our picnics and our sky-musings.

TO RENEW

Sometimes it gets to be too much. Sometimes I can't catch a break from all the obligations that pull at me. Doing all of the perfunctory domestic things, from paying bills to cooking and cleaning, and caring for the kids—all the running around and sign-ups and planning—trying in vain to get them to listen, quelling their sibling arguments, and then caring for the dog and stepping outdoors to chickens and a turkey always wanting a treat or attention, and a garden needing watered, needing weeded or plucked or pruned. The constant needing—the hallmark of motherhood. I can't tell if I am trying to do too much or if there is just too much to do. Do I overextend myself or is this just how it is given this stage of my life?

We'd talked about getting miniature goats for a time and the kids still press, asking when we're building our goat encloser and staking claims on what to name them. I'd love to have them too, especially to clear out the woods. Those things will eat poison ivy, barbs, black snakeroot, and whatever else is most unwanted. But they're also escape artists and loud and smelly and needy and I just don't need another thing to do. Nurture what we have and all. I'm learning I need to scale back.

With all our home remodeling projects completed, I find myself pointing out things to change, like the carpeting on the stairs that I should have never selected, stained now with dog paws and tracks from kids that don't take off their shoes after traipsing through the woods. But new carpet will only get ruined again. Dirt, dog hair, a never-ending litany of crumbs left behind by the husband and the kids as if every day they needed to Hansel and Gretel their way through the house. Paint dulled by handprints and never a view window-wise without bird droppings or nose smudge. For all of our hard work, this home is still scratched and dented like an "As-Is" used car at the dealership. In other words: no guarantees.

Around this time of my broody, exhaustive to-do list, I posted some patio furniture we had no need for on Craigslist. As with most things I've listed there, it was gone in a day. But it seemed that this badass woman who came for it was sent from the universe to deliver the words I needed to hear. Often, I can't wait for the buyer to load and leave, double checking the locks at night and kicking myself for bringing a potential threat to our property. Yet every so often, as was the case with her, I make some interesting connections. She was going to sand blast and repaint the furniture as a gift for her pregnant daughter to enjoy rocking the new baby from their back porch.

Her husband thought she was ridiculous for not just buying new and being done with it, but she was adamant that it was better for the environment this way, to reuse. Besides that, she liked the accomplishment of doing the work, and believed the gift would mean more if she had a hand in it. We stood in the driveway for over an hour talking kids, politics, life, the interesting encounters we've had via Craigslist, and when she was pulling away with her loaded trailer, late for a dinner meeting, she paused, hooked her arm through the open window of her SUV to crane her neck around and impart some final words: *Be gentle on yourself.*

I've come to see how important that is—to be gentle on myself. It means I can't do it all and I shouldn't either. It means instead of getting frustrated cleaning the same mess made by the kids or the animals or Matt, to see it as life being lived in. On hold with the cable company for 40 minutes with a mind racing north and south and east and west with all the things needing remembered or done, and feeling like I can't clear the fog; the only answer is to care less about it. So what if I sign the kids up late for x and y and the house isn't sparkling and dinner is cold. Like the garden, the answer is in scaling back.

It's akin to what my mother once said about raising me: *Casey, we had no idea what we were doing and just did the best we could with what we knew.* To admit that kind of vulnerability is a profound call to self-love. I don't always get things right, but I do have good intentions and do what I can with what I have available to me—try to plant and nurture not just this garden in the ground but the garden beyond—the kids, the marriage, the birds, and the space we call ours—and that's not nothing.

TO GROW

We redid the garden. Years of pumpkin vines buckling the
flimsy fencing and a milieu of storms knocking at the posts
that we hadn't set with concrete had left it shabby. This
time we used larger, proper timbers for posts and secured
them two feet in the ground with a concrete base. We used
a higher gage of fencing and secured this too with treated
2x4s along the top to keep the fence from wobbling. And
because our years of working compost into the soil still
didn't inspire better growing, we built an assortment of
raised beds with whatever scraps of wood we had around
and filled these with primo grade dirt.

The old garden had served us, but we had learned to
build better. It took tinkering and trying and failing to
figure this out and maybe if we had done more research
instead of jumping right in with our excitable impatience
we wouldn't have had to rebuild, but there are things we
need to come to organically, in time. Better wisdom in
the experiment of trying.

We ordered a truck load of pea gravel and rolls of landscape fabric to cover the ground around the raised beds
not just for aesthetics but to function as a weed blocker
and free us of extra weeding work. For all of that, some
things still refuse to die. Amazing that we coax dirt, line
plantings up for optimum lighting, add water and prune
and de-bug all to garner a bit of harvest when there are

so many unwanted weeds that sprout and flourish in any condition—under fabric and rock, in any nook, on a speck of dust with a lone dragonfly tear for moisture—we have creeping charley, burdock, and snakeroot that find a way to thrive. Dandelions spread like a chorus of bodies mid-wave in the grandstands. Prickly thistle undiscovered until you walk barefoot. Plantain, stickseed, horseweed and the immeasurable unnamed.

Why is it so much harder to grow the garden we intend than it is for a weed to just grow? It takes just a wind or a passing bird dropping seeds or the dog's fur to plant the weeds. And then they colonize underground, their armies fit to tackle even the most hostile soil conditions. They just become, with a bit of air. Perhaps it's Mother Nature's way of reminding us that nothing is certain and nothing is owned. They become by chance.

In the margins of my ancestral tree—an actual notebook written by a distant relative on my father's side, their handwritten accounts of stories passed down, recorded, photocopied, shared—I'm struck by the name Nora Lane Hickman. It wouldn't have been her birth name and there's no way to tell when or where she was born, but she existed with a kind of chance and resilience the luck of timing is keen to outlast. Her story—a series of happenstance.

It's written that "Uncle Cleo" and "Aunt Mabel Oxenreider" stayed with Nora one night in the last two weeks before she died and she told them she was born in the tribe of Crazy Horse. Earlier mentions of her stated she was "full-blooded Indian," either Navajo or Sioux, but I'll go with Nora's remembering. She told of pony soldiers (calvary that adopted the ways of the Native Americans in order to better fight them) that surrounded their camp one night: *Burning everything, killing everyone and throwing the kids on the fire.* She and her mother escaped the massacre and kept hidden in the brush, traveling from that stretch of Kentucky by night, subsisting on whatever they could find. She was three or four at the time so her account and the record of the piggyback of words are sparse.

The next line reads: *Her mother was shot and died,* and the next: *Some people came upon her and gave her a ride,* their wagon crossing through St. Joseph, Missouri and up to Iowa, where the Henry Lane family took her in and raised her. More words are written about how Henry Lane got the girl a birth certificate than how she came to be with the family. It recounts how he went to a doctor to request one for what he said was their new baby girl, how he told the doctor she was not yet named, whereby the doctor signed it and instructed Henry to take it home and send it in with $1.00 when they came up with the girl's name. When Nora's name was finally registered, her age was *6 to 8 years off.*

Life is a string of unending "If-not-for's." If not for her mother's quick secreting-away, if not for the travelers who offered safe passage and of course Henry Lane's family for raising her as their own—imagine how different the timeline her survival would root and grow. A string of ancestors nonexistent. The smallest thing can alter the path we come to know. It takes work to make it here, sure, but luck also has a potent role. A series of decisions not entirely our own, as if the future were dependent on the precariousness of wind. We spend so much energy trying to tempt this plant to grow or any number of actions to yield our hopes, if even that is to just stay afloat, all in the face of such uncertainty, our fate in the hands of strangers—what a remarkable thing is our survival. Like a plantain, seed-born from a dog's stubborn paw, thriving in the cracked, well-used blacktop, finding a way to sing still.

Still, we've grown multitudes here and in ways beyond that plot of land we sow. It's made us better observers, more patient, and sensible. We celebrate what luck we do have. Turns out we make an excellent blackberry and sage jam and the hodgepodge of tomatoes we pick, when processed and canned, make for the best red sauce base. Jarred pickles, pickled green beans, pickled peppers, all of it tucked with our fresh herbs and garlic, in rows and gleaming from our basement shelves.

A Tooth in the World

Friday night and the dogs take
to the first fine hour before the sun
cashes in to belly their backs
on the grass as if to scrub
the essence of the grubs some
two feet under.
If they could get any closer
to the gut of ground they would.

I read once that a dog's sense
of smell is so keen, when given
a bone they know not just
the animal it once hosted, but
the creature's last meal and how it died
and whether or not they feared
it as if the level of trauma
one endures steeples
at the very marrow like a choir
of forget-me-nots come dawn.

Twice, I've given birth and twice
I was pure blood and bone
and push with a tooth in the world,
completely nothing
more than it.

My old dog believed he did life
a favor fetching a ball—he'd take
to running before my arm even
made its' turn, trusting
it would show. Love
is always sending that ball flying.

THE FLOCK

Mento died in a freak accident. Fisher had placed his 1961 Chevy Apache bench seat in the back yard (one that a neighbor had given him to refurbish and if successful, would gift him the whole truck, piece by piece, to restore and eventually own). He's an amazing kid—always befriending neighbors, especially older ones with more stories to share. At 11, when most boys are playing video games or sports or on electronics, he is outdoors digging for antique bottles in old dump pits or else creek-walking for arrowheads or else offering to mow, shovel, rake, or doing any odd job needed.

He'd spent the afternoon removing rust from the bench springs with steel wool and 100 concoctions that might work and then left it teetering on the slope out back while he went in for dinner. We heard some squawking, but that was ordinary. It wasn't until later when counting the birds at lockdown and one was amiss that we saw poor Mento had been crushed beneath the bench. Thumper had taken to perching on the seat all afternoon so we could put two and two together. Fisher was awash in guilt and grief. A hard lesson to learn that despite our best intentions, sometimes terrible accidents occur. Yet it's an important one for a kid to know—think ahead and think of consequences. And dang if kids don't end up facing that one again and again.

Another of our quarantine chickens (the Amberlink) was killed via owl one afternoon in broad daylight in our woods. She was a loner and a good forager who'd wandered close to the creek under an old oak tree when it hit, too quick for Luna to get there in time. I came upon it right after as it winged away, my chicken's head ripped clean and dangling from its sharp beak. Olive, too, brought worry. Because I knew her mannerisms well, when she stopped hanging about with Thumper and greeting us when we set foot outdoors, appearing weak and uninterested in her favorite treats, I knew something was off and searched up her ailments. A common cause for her condition was being egg-bound, which is when a hen has an egg stuck inside them that they can't seem to lay, a situation that is deadly. The recommendation was a massage and a warm bath. Phoenix and I bathed her in our newly remodeled bathroom, wrapped her in a towel and sat petting her awhile. She thanked us by laying a shell-less egg on our laps. Not only was she not egg-bound, but apparently not getting enough calcium in her diet. We already offered the chickens a wide range of food, including crushed oyster shells which offer lots of calcium, so I hand fed her a bowl of another recommended item on the list of calcium-rich foods to offer chickens—plain yogurt. After a few rounds of her shell-less eggs and our yogurt hand feeding, she began laying eggs with regular shells on the regular.

It's wild to think what I've missed. So many times, I've happened by the kitchen window and there's a car driving slowly, turning (thankfully) to miss Thumper. And Thumper does the strangest thing when a car gets close or danger is eminent—she just lays down right there on the spot as if resigned to her fate or perhaps believing the best of the world. Or because she's a bird and doesn't know what a car means. Backing out of our garage takes one kid to clear the area of birds and returning takes another to exit the car and retrieve Thumper, who, having run to greet us, plops herself right down in my approach. I can't imagine the times I haven't been here to witness.

Winter gives a reprieve from the worry as the birds don't like the snow. They spend much of the season cozied in the coop and Thumper lays more eggs then too. Her speckled eggs are twice the size of a standard chicken egg and twice as good—richer in flavor and a denser, deeper golden yolk. The coop stays locked up and the birds hunker in, away from the frigid winds. I still tend to them daily to ensure they have fresh water and food and to gather eggs. One January day, I'd just returned to the house from doing so when I heard a knock at the front door. I opened it to Thumper standing there, a couple of chickens flanking her sides. She yelped and turned, looking at me to follow.

Apparently, I forgot to latch the coop door when I refreshed their food and water, giving them a chance to get out before the wind blew the door shut again. They couldn't get back in. It stuns me to think that Thumper has the wherewithal to march her buddies through the worn dog tracks in the snow around the house to the front door and to know to knock on it with her beak so as to make some noise and find me. And not only that, to make like I needed to follow, which I did, along the same worn snow path around the house to the backyard coop to save those chickens.

She knew to come to the front door and not the back door attached to the garage, where I likely wouldn't have heard her knocking. I often wonder if she tried that door first, waiting a beat before trying the front. Something tells me she didn't. Either way the fact that she knew how to find me, knew to knock to get my attention à la old Lassie, pretty well blows my mind and makes me never want to see a turkey trimmed and dressed as a centerpiece on a Thanksgiving table. It also explains how turkeys have been around for some 10 million years, surviving two major threats of extinction. I sense she's more of a mystery to me than I am to her.

This isn't the only time she has protected the chickens either. Every thunderstorm, I find her blocking the entrance of the enclosed coop to the fenced run so that the chickens won't fate the wind and rain. And when broody, she steals their eggs as if to hatch one for them too. From the occasional droplet of egg yolk on her beak, I've reckoned that she gathers every freshly laid egg in every nesting box we have and carries them in her mouth to form the pile that she'll sit and brood upon. Turkey-chicken crossbreeds are a real thing, and real ugly. They call them Churks and they only produce male eggs. Of course, Thumper was too large and too aware for the rooster to attempt mating, but still I'd let her sit broody, believing in time she'd hatch a bird so at least for a time I could worry less about her wandering the neighborhood, testing roads.

I've had neighbors call—*Turkey over here in my driveway!* I've had Matt come home with the news that he passed Thumper on his way, down the block resting in the grassy corner not two feet from the bustling road. I've stopped counting the number of times I've had to hustle through the streets to retrieve that turkey, looping my arms around her and resting the bulk of her on my hip for the trek home. Cars will slow and gape, the drivers rolling down their windows to inquire if I was indeed lugging around a wild turkey. *Nope*, I tell them, *just my naughty pet.*

So, when I'm not spending every evening searching for and coaxing that turkey, worried for her safety, her being broody does me fine.

TO BLEND

One gift in the early days of the pandemic was to have Ella home. The kids grew closer than they would have otherwise, bonding over games and music and memes— their way of coping with their deep, unnamed, and new feelings of uncertainty as masks came on and the world they knew shuttered. It was hard, but there were things we loved about it. Mostly the time to slow down and enjoy what we had as we were no longer caught up in the endless run around of activities and going-ons.

Fisher spent even more time out in the woods and inquiring from neighbors whether or not they had an old dump pit he could dig in or what they knew about the area's history. He intended to write a history book on Dogpatch, and dug up some interesting facts, including that much of the area was owned by Edwin T. Meredith, who raised cattle on the spot while working as an entrepreneur. In 1902 he founded the Meredith Corporation, a large media conglomerate, and in 1920 was appointed as US Secretary of Agriculture under Woodrow Wilson. Not far from our home, one can still find the Meredith mansion, which was built in 1939 to entertain service men and women during World War II.

I caught the boy attempting to skin a raccoon that Luna had killed near the chicken coop, even though I expressly forbid him from doing so when he asked. The stench and blood stains on his hands convinced him it was a bad idea—he, who had high hopes of making a coon skin cap the likes of Davey Crockett—ended up burying the unfortunate intruder in our raccoon cemetery and took to tracking animals instead, one of which ended up being a coyote.

Great Pyrenees can easily take down a single coyote but where there was one, there was likely a pack. Luna needed backup. Craigslist, another farm visit and time to raise a puppy while we were mostly still home. The farm where we found him was so flat and rural you could see miles in any direction. The guy had an assortment of Pyrenees as well as ducks, turkeys, chickens, and a Jersey cow that could have taken off my head when he jumped over me as I knelt in the barn to see the pups. We brought Blue home when he was just six-weeks old. He was born on the coldest day of winter while the farmer was away and all but two puppies froze to death. It's unheard of to get a puppy so young, but farmers have a different kind of attitude towards their work dogs—not unkind, just that they are not the same as a city house dog.

Pyrenees are known to be good guardian dogs, yes, but they tend to be stubborn, aloof, and slow to train because of it. As it turns out, Blue is the quintessential example of all of those traits. Thumper, inexplicably and because she has more trust in the world than she should, gets too close to him. So far, he regards her as an intrigue and not a chew toy. Even though this breed is known for protecting livestock, they have to be acclimated and trained. You can't toss a puppy in a coop with chickens and not expect them to play rough. It comes in time, and I don't let Blue out when the chickens are free ranging unless I'm there to keep watch. Luna can sit and guard the property for hours when we're away, never testing the road or the neighbor's line. But when the kids and I ran to the store one cloudy late morning, we put her indoors, mostly to keep Blue from chewing on furniture as he was apt to do when left alone. We were gone not even an hour.

THE FLOCK

A gathering of feathers first, the pile of them belonging to a Barred Rock, the remainder of our old rooster's lineage, nestled in a circle astride the driveway when the kids and I returned from the store. Now chickens lose feathers. They molt, they drop them when startled and I thought nothing of it when I saw Bang Bang perched on the lip of my flowerbox a few feet away. But then I noticed the quiet. How Bang Bang seemed frozen still, how Olive didn't dash out to greet me as always and that quiet grew large. I rounded the house and headed straight to check the coop where I found Thumper, alive and broody in the nesting box along with our Black Sexlink, Stoney, but no other chickens in sight.

We let the dogs out and immediately knew from Luna's frantic nose-to-the-ground whining, zigzagging all over the yard following a scent that some animal had happened. I found the body of the driveway feathers on the path to the creek, 100 feet from where they'd fallen. For the next few hours Fisher and I, with Luna, would scour the woods, snagged by stickweed, brushing poison ivy, stuck with cockleburs, hollering and panting, searching for the eight other missing chickens. We found our Buff Orpington, Prince, submerged in the creek's bend, one wing missing from the struggle. We found the one remaining rooster-hatched mixed breed chicken stiff under a bramble of branches at the back edge of our property line.

We found clutches of feathers, knowing exactly which bird matched them all over the property like some dot-to-dot we were trying to put together to reveal the larger picture it showed. We took to the back bike trail and spotted a few more—Thumper's best pal, Lucy IV, likely gone. Not a lick of Olive. Miraculously, after hours of searching, Frenchie appeared in the neighbor's yard, clucking, startled, but alive after escaping yet another attack. You wonder why the rest didn't see fit to jump the fence and hide, to find high ground. Five bodies were missing and three found dead. With those kind of numbers it had to have been a pack.

Fisher and I tracked coyote prints in the soft mud of the creek and he, old soul, visited with a neighbor who confirmed he'd seen a few coyotes that day stalking across his back acre. You want to hate those wild dogs but they live here too. This land is just as much theirs and at the end of the day, we're all just passing through. What can we do but learn and grow from it, hope, and do better still. Not many tears have been shed for the countless losses over the years, aside from that first, as the flocks came and went and our connection to them dwindled, but this group hit square in the chest where the heart does its singing. There's a bit of numbness that follows, a bit of kicking like it were a rock in the way at the wonder of why these chickens couldn't find safety. They can fly a bit, so why didn't they fly to higher ground to get away from their assailants? Why didn't they follow Frenchie, with her hard-earned foresight to jump the fence and hide in the neighbor's woods?

The hardest thing is keeping Thumper and the chickens caged. They want to forage, to splash in the cool dirt on the hillside, to roam the grounds and stretch their wings—to free range. It's not natural to be caged and frankly, Thumper deserves to live her best life. Yet I can't always control the harm in the world and I want to keep them safe so it's hard to reconcile—how does one explain to a bird that caging them is for their own good when really it's for my peace of mind? It's a lot like having kids—you know they need to explore the world in order to enrich and understand themselves. You know they're not yours forever, but you are always saddled with worry for their safety, wanting to hold on to what bit of time you're given for as long as you can. At a certain point we all must let go, come what may. All we can really do is teach them well and send them out, hope the world is kind to them.

So here's what you do—keep them locked in, safe for a time. You forgive the coyotes for their instincts and let the dogs out longer that their presence might send the predators another direction, much farther away. In fact, you frequent the outdoors even more, passing the buried remains of your favorite flocks, only letting the birds out to free range when you know you'll be there to keep watch, which you do time and again. And then bit by bit, the edges blur. Uncertainty will always be uncertain and no amount of fencing or dogs' watching or wishing will draw the conclusion. We try our best to be sure, doing our part based on what we know and what we're given to make it so it's easier on them, but there comes a time where you have to let things grow where they are.

I Love You Because

I watched you kill the chicken that needed
put down after we gauged she was too far
gone to live dignified—how you held her
before you brought the knife to her neck,
the assurance you spoke then, painting
an afterlife for birds and believing it—
and how you calmed the twitch of her
body afterwards, kneeling with her remains,

your hand asking not forgiveness but space
for mercy. Because you didn't want to have to
but you did as any alternative meant prolonged
suffering, but yours meant this—quick, done.
Because you dug her grave and asked if I
was okay—that. And because you were not.
That you did what you didn't want to so gently
like a leaf at Autumn's end, the worry your own.

TO BLEND

Sage words my father once gifted 16-year-old me—*Sis, this is just an inch of your life*, and it was, time and space assured. Truth be told, I hated my life then. Couldn't find a way to fit, and I didn't want to. I wanted nothing to do with the Friday night football games, the cliques, the small-town offerings, wanted nothing of the looks and advances from older boys as my body developed and developed early—bait for their hormones, and bait, too, for girls to take their insecurities out on. I became numb, indifferent, took to reading philosophy and cutting my forearms with razors late in the night as a means to escape. Here's the mind of a cutter—this will release a bit of the abyss of pain that saps bone-deep.

As for reading *Being and Nothingness* by Jean-Paul Sartre, or Milton or Nietzsche or Thoreau at that age, I was trying to locate some meaning, to give shape and legs to the story I seemed too stuck in. A kind of hope to muscle towards. Because on some of those Friday nights at the football game, girls would spit their gum in my hair, or rumors swirled, and some guys took to calling me "tits"—the needle and the voodoo doll. But hearing that line—*this is just an inch*—well it got me through. One day, this would all end and I'd be looking back at it and not through, *c'est la vie*. All of us on a timeline, the give and take unknown.

So when I ask the kids what strikes them most about our start here—what to include in this telling of this stretch of time in our trajectories, their answers surprise. They all mention the time we built a teepee from felled logs by the creek, which still stands, or painting tree trunks at random, the paint still clinging on. Phoenix: *How we explored the woods like it was a new place every day and looked up at the birds in the trees and just watched the sky.* Fisher's best recall is the fort he dug on a steep slope in the woods—no sense of erosion concerns—four feet deep and wide complete with wall bracings and a camouflage net so that anyone without knowledge of his spot might break a leg. Secreted shovels, stashes of firecrackers so that he and his friends could play WWII, complete with thrifted uniforms and pellet guns. A total eyesore and bad news for the hill come rainstorm; it's filled in and forgotten now. Elle remembers most the time we purchased a huge jar of pickled sausages—the kind set to float on a shelf next to pickled eggs and whiskey bottles at a dive bar that you assume have been there for a decade because who in their right mind eats that? A birthday gift for the kind of guy that does: Matt. The kids and I lugged that jar throughout downtown, taking pictures with it like some long-lost twin, on the train tracks, on some random bench, at the bridge where he'd proposed to me.

All these little blips I wouldn't have thought to include turns out formed their reminiscing. The weight of things carried, the seeds planted, the intentions—it's a wonder what grows. Elle did not spend half her childhood here, but what she spent seems to have sowed the kind of future she hopes. She says after she earns her degree in cyber security and Russian at the university, she just wants some land with a big dog and garden.

Time and space. When I tell my father now how much those early words turned out to mean his brow line arrows. Older now, so that inch, the assortment of them, have become a gift and also a matter of how long, he says, *Can you believe our lives are just an inch on a line, a dot with a name on some future ancestral tree?* A dot, and all those days we wished would end. He says one day he'll just become a story for the next generation and wonders aloud how long before that story ceases to be told. So long as I'm here, Pops. This season, the next?

TO RENEW

We plan to tear down the garden to make room for a good-sized workshop so that Matt can follow his passion to start a coffee roasting business. We'll still garden, to be sure, just elsewhere on the property—the nooks and crannies where we've seen the best light, not in line with the sharpest wind or the trails the deer seem to favor. I've taken a new job at the school with good insurance so that he can regroup, and together we can start this new venture. He learned the trade back in Alaska working as the master roaster for North Pole Coffee. He started his own roastery in Minnesota and cofounded the Midwest Roaster Guild, where he stood in as president before moving down here to raise Ella, so it's always been at the periphery of who he is and what he wants to do. I believe we ought to do what drives our hearts' intention, so I'm all in. We're going to name it Bluestem Coffee as bluestem, a kind of homage to the Midwest—our roots—is the most common type of prairie grass here.

Now we know the best light. We know to build raised beds and fill them with well-fertilized soil from our chicken manure compost and use sturdy, treated posts concreted in and over six feet high for our fence to keep the deer from treading. We know we'll plant tomatoes, butternut squash, cucumbers, peppers, green beans, potatoes and herbs and microgreens, forage our berries and rogue mushrooms, pluck apples and pears. The basement shelves won't be lined with a dozen varieties of canned goods, but what it will have is loads of jars filled with my blackberry sage jam, stewed tomatoes, pickles, peppers and green beans, boxes of butternut and honey nut squash in the cool, dry corner. The freezer stocked with chopped herbs preserved with olive oil to melt into the soups we'll make all winter. The rest of our goods eaten in season.

It's not like we failed, but rather discovered via trial and error the way for us, and even giving up at times means this—we tried. Over the years I've grown wide-eyed with a number of flash-in-the-pan interests, believing at each turn this new foray would weave itself with a kind of meaning in my life, from knitting to cheese and candle making, to running a marathon or learning to play an accordion, making silk-screen one-of-a-kind shirts, and not a one of them now more than a blip. None of it stuck. But I can think of no better love letter to this life than to be interested in the possibilities and care enough about the timeline to try. To come away with a bit of knowledge, a finer tuning in this evolution of heart, as if to say, *I did something here*, as if to say, *perhaps*.

There are 11 chickens and one turkey free ranging the yard now, the threat of coyotes, hawks, coons, and minx always near, but a fortified coop they'll roost in come dusk and two dogs that guard the perimeter. For now their chirps and quirks and eggs are ours, but I can't always keep them from harm. Perhaps they'll persist, perhaps they won't. In the meantime, I offer them space to roam and perch and an assortment of all their favorites foods. I do what I can to give them the best life while they're here.

Strange that we start from zero—unaware of life's borders and markings, learning as we go to persevere, give meaning a shape that is definable, only to come to the knowledge that it takes a kind of letting go to flourish. It's a lot like gardening, which takes a kind of whittling down, a lot of waiting, the luck of weather and still no guarantees. There comes a time when you have to cut your losses, praise the bit of victory that comes. Contentment, the elusive happiness, is in those small moments too. For me that means dew drops in the leafy palm of a Brussel sprout, locusts and toads and fireflies exploding the twilight, bonfire, the moth on the window whose wings shape the perfect heart, Thumper answering her whereabouts with her high-pitched click from the woods, the mouth-burst of the season's first cherry tomatoes, a quilt spread atop my favorite hill come sunset and the husband and kids that share it, or when Elle, home late, kisses my sleeping forehead, when Phoenix drops an *I Love You, Mom* at random and when Fisher, the young teen of him now, rests his head on my lap come family movie night, also Matt at 2 a.m. watering the garden or making a mess of the kitchen prepping his late meal after having put us all to bed with soft words and the mystery of his night I'll wake up to, early enough to catch stars, the aura of sunrise, a kind of poetry that makes anchor in my ribs all day.

I call that lucky.

TO GROW

To have an urban farm is to face a maelstrom of failings. Chickens die in a number of ways. Sometimes a hailstorm barrels through, decimating crops, flattening stalk and stem. Sometimes a harvest comes when you least expect it or it hardly comes at all. And too, the hiccups in remodeling this home and stressing over kids, finances, health, broken relationships, and on, you'd think I'd be hardened, but I am not. This ebb and flow requires softness to flow through the sharp edges or else we'd be stuck repeating them anon, moving nowhere. There's a bravery in accepting change and not just moving on but making something else of it. You learn to refine and push forward; you learn to grow best what best you can.

Divorced, uncertain of anything, I discovered a path that I never would have found without that breaking, one that this universe was waiting to deliver in time. I wouldn't want anything to change that. Perhaps the multi-universe exists and there are innumerable versions of me out there traversing different fates, but this one here is mine.

Matt's disease has been held back at the fringe like the woods along this road. It can flare up at any time, but such is anything. In the meantime, we sit on my hill in the evenings, walk barefoot in the grass, lucky for moment. Our kids make a thousand messes—random assortment of shoes strewn about the halls, wrappers and cans stuffed under the couch, crumbs all over the counter that I sweep off time and again, but this house is lived in. Our kids are kind. By now the improvements we made on our home could use some refreshing—kids, dogs, weather, and time have worn it down, from the stained carpet to scratches on the wood floors, knicks in the walls, and the to-do list never finished, always, like a leaky faucet, adrip with more. Keep planning for the next project and then the next.

The difference now is me. It'll come, in time, and not by way of complacency, but because I know I can't do it all. Because not everything is meant to be no matter how hard I try to make it so, because time is too short to be so hard on ourselves. To be here at all on this planet some 44 times now in its path around the sun is humbling and miraculous. That I have breathed and grown and given words and hope and love is not nothing. One bit at a time, seed by seed, root by root. Why not sit back now and then and be here in the presence of what I have done and what came true of this bone-tired, blister-worn dreaming? It's a kind of grace I give myself now.

Though nothing has slowed, my energy to do them has. My motivation too. I've never been one to procrastinate but lately I've been embracing the idea that I don't have to do it all, at least simultaneously and *right* now. Some things can wait. The bathroom ceiling has started to peel and part of me thought it would take nothing but an afternoon to fix it. A month has gone and all I managed was to scrape away the flakes and apply three coats of sheetrock mud, sanding in between their supposed quick-drying time. I need to paint over this wee bit of ceiling with a layer of sheetrock paint and two more of satin enamel, but the thing is I know it will mean painting the whole ceiling again to match and the other thing is this—that ceiling isn't going anywhere because we aren't, despite whatever inconveniences and what-ifs house wise gain footing. This land is our "here is good" soliloquy.

Matt, did we ever hard-knuckle our way to this, and could we still? It's hard to tell, what with our bodies in different places, the stretch of morning leaning longer on the windowsill. It was easier then, but something tells me we'd find a way to make anything work. Whenever and however, we are destined to *home*.

The only certainty is that nothing is finished and we have no idea what will become of any of it. To love it here is to accept impermanence and do it all anyway.

Sometimes Forgetting

Seven months pregnant in Lamaze class,
the instructor had each of us hold an ice cube
in our hand and squeeze until the cold sting
became unbearable, but later,
instead of dropping the melt from
our palms we'd be tasked with braving
the stabbing harbor beyond our edges.
And don't worry, she said, the shock of pain
your body endures will be forgotten in a few
months time—how else would anyone consent
to deliver again? As if our species
could depend on our forgetting.

But that amnesia hormone—a myth, and most
remember the stabbing burn
years on. Still, I like the story of erasure
for seeing new—the backyard tree that turns
goldenrod come October, giving pause
to a day wrought with tedious work, or the drum
solo of a full winter moon like a pledge to do better
as if nothing else did sing till this.
And how often I tell my kids that despite the grip
of aches, tomorrow is new because it is even
if it isn't. Perhaps it's hope that needs
a kind of forgetting, again and again,
to be had.

EPILOGUE

Fifteen minutes from home and my phone buzzes with a text from a neighbor, which I read red-light quick: *Thumper was in our yard so I carried her back to yours.* School is back on for the kids so work for me is a solid 15 hours of my own combined with driving kids to where they need to be and running errands, cooking dinner, planning, prepping, waiting. I get home with a car of groceries to unload, a meal to start, a husband to say hello to, and at the periphery—the notion to check the yard for the turkey who I'd let out an hour before because she saw me and clucked, pacing the birds' cage as if to say she'd been caged enough, as if to say, *Let me happen*, and so I did, because I know a thing or two about wanting to be free of what surrounds me. I dropped the grocery bags from the car and set out to make sure Thumper was still in the corral of our yard. There was no sign of her and no chirp in reply when I hollered.

At the curve of the road leading to the neighbor's house, where a smattering of trees edge, where I look from my favorite perch of hill—sunset, stars, the road spindling west—there laid the lump of Thumper. And because we don't always want to believe what we see, I ran to it hoping a high branch had fallen there and made a dark splotch of the road, even knowing there lied my sweet Thumper, her body still and warm.

So many what-ifs follow—if I'd checked on her before unloading the car, if I hadn't let her out yet, and the neighbor's own: if she hadn't carried her back home—but to house guilt or wonder doesn't solve anything and sometimes things in life can't be solved but rather felt and that's exactly what happened. Fisher did most of the work digging her grave while Matt took Phoenix to dance class and I stood, shovel in hand, sobbing and lamenting—*why dammit*—even though I already knew why means putting it all out there, why means putting faith in things to happen knowing they won't happen always.

And all that evening I sat crying about Thumper, I kept hearing her sound the woods as if answering the something she would. Little chirps here and there, to the point I asked Fisher if she was indeed what we buried—enough of a resemblance that the sound beckoned me back to the road a bit later to make sure she wasn't sauntering home yet, as if we'd just buried some wild twin. Instead, squirrels uttering her tone identically, and a distant neighbor jogging by, riling the dogs. I'm not sure I can ever raise another turkey and certainly none could replace her. Still, I'm thankful for the intrigue she brought to my days and still, she amazes that she mimicked the sound of local squirrels, toads, and birds to the point that to hear them now is to hear her. I cannot change the outcome, but I do not regret having her here, however long, because my heart is better for her. My awe for nature running deeper than it would have known.

I like to think she knew what she meant to me and perhaps the memory of us side by side in the grass fixed in her mind before she died as if she too had a moment flash before her but I know this life is one part what we wish for and one part an amalgam of everything else—the wrought iron of experience, the nuance of happenstance. Still, hope is a real thing. It's why we build and plant and love otherwise what would be the point of anything.

The world is quiet for a time when you lose something revered. A kind of echo remains, as if the days that spill before us were ombre—half here, half song, remembering. Meanwhile the pear tree we planted jewels sweetly for weeks, and what the bees don't mine, we do. The blackberry bushes burst, stain, fill our jars. Autumn comes in splashing, then hushes bare. What beauty there is, and despite the work and time and setbacks, the losses and the wonder, what luck it is to be here to witness, to set foot in it at all.

ACKNOWLEDGMENTS

Thank you to the publishers in which these poems, sometimes in earlier versions, first appeared:

"Coming Home" from *Ground Work* (Main Street Rag, 2018)
Chiron Review: "Note to Future Self"
Cold Mountain Review: "Pecking Order"
The New Territory: "Old and in the Way"
　*2021 Pushcart Prize nominee
Rumble Fish Quarterly: "A Refining"
The Westchester Review: "Forgive Everyone Everything"

Thank you to Cornerstone Press: Director and Publisher Dr. Ross K. Tangedal, Production Editor Mattie Ruona, Editorial Director Ellie Atkinson, Managing Editor Chloe Cieszynski, Media Director Ava Willet, Natalie Reiter and Sophie McPherson in sales, and the entire staff.

Special thanks to Phoenix, Elle, and Fisher for inspiring not just the stories herein, but to a greater extent, my spirit. Thanks to John Domini for nudging me to start penning my varied backyard chicken woes. And to the incredible pack of women and writers I'm blessed to call friends—Emma Murray, Jessica Gunderson, Kristina Lilleberg, Gwen Hart, and Cassandra Labairon—thank you for your encouragement and for helping me define my words. And, of course, to Matt Knott—my person, my ally, my home. I love you all.

CASEY KNOTT is is the author of *Ground Work* (2018), a poetry collection, and the poetry editor for *The Wax Paper*. Her work has appeared in *Gulf Stream, Prism Review, Cold Mountain Review, The New Territory, Cimarron Review, Sugar House Review,* and others. She lives on an urban farm in Des Moines, Iowa.

Milton Keynes UK
Ingram Content Group UK Ltd.
UKHW040713060824
1160UKWH00007B/365

9 781960 329493